W9-BXQ-790

PHYSICAL BIOCHEMISTRY

K.E. VAN HOLDE

American Cancer Society
Research Professor of Biophysics
Oregon State University

PHYSICAL BIOCHEMISTRY

SECOND EDITION

PRENTICE HALL, Englewood Cliffs, New Jersey 07632

Library of Congress Cataloging in Publication Data

van Holde, K. E. (Kensal Edward), date
    Physical biochemistry.

    Bibliography: p.
    Includes index.
    1. Biophysics. 2. Biological chemistry. 3. Chemistry.
Physical organic. I. Title.
QH505.V27 1985   574.19'283   84-8266
ISBN 0-13-666272-2

 © 1985, 1971 by Prentice-Hall, Inc.
A Simon & Schuster Company
Englewood Cliffs, New Jersey 07632

All rights reserved. No part of this book may be reproduced
in any form or by any means without permission in writing
from the publisher.

Printed in the United States of America

10 9 8

ISBN 0-13-666272-2

Editorial/production: Nicholas C. Romanelli
Manufacturing buyer: John Hall
Cover design: Ben Santora

Prentice-Hall International (UK) Limited, London
Prentice-Hall of Australia Pty. Limited, Sydney
Prentice-Hall Canada Inc., Toronto
Prentice-Hall Hispanoamericana, S.A., Mexico
Prentice-Hall of India Private Limited, New Delhi
Prentice-Hall of Japan, Inc., Tokyo
Simon & Schuster Asia Pte. Ltd., Singapore
Editora Prentice-Hall do Brasil, Ltda., Rio de Janeiro

574.19283
V256p
1985
c.2

To my wife and children

CONTENTS

PREFACE

Since the publication of the first edition of this book some methods that were widely used are now obsolete and new techniques have emerged to dominate the field. An obvious example of such change is electrophoresis: moving-boundary electrophoresis has virtually disappeared as a reseach method. On the other hand, gel electrophoresis has become a major tool in molecular biology. Similar examples abound. This book has been rewritten to reflect the many changes and developments in physical biochemistry during the last decade.

Because there is need for a concise, readable introduction to the methods of physical biochemistry I have resisted the temptation to make the book longer or more difficult. In a few places new theoretical material has been added (the treatment of cooperative binding is an example) and some derivations have been made a bit more rigorous. However I believe these changes are appropriate since in my opinion biochemistry students are now better prepared in mathematics than they were a decade ago. Most of the additions have been balanced by the deletion of material now considered to be of less importance.

Many new problems have been included and, in response to numerous requests, answers are provided for half of them. The remaining problems are left without answers, for instructors who wish to assign such problems. I express my appreciation to the many individuals who have contributed ideas for the improvement of the book. The assistance and helpful suggestions of my wife, Barbara, during the preparation of the manuscript were invaluable.

K. E. VAN HOLDE

PHYSICAL BIOCHEMISTRY

INTRODUCTION

Cell and tissue, shell and bone, leaf and flower are so many portions of matter, and it is in obedience to the laws of physics that their particles have been moved, moulded, and conformed.

D'ARCY THOMPSON, 1917†

These words tell what molecular biology is all about. By proceeding from the assumption that the fundamental physical and chemical mechanisms of life are knowable, it has become possible to understand, at least in part, how inheritance works, how cells make proteins, how proteins regulate metabolism, how muscles contract, and much more. Because the fusion of physics, chemistry, and biology we call molecular biology has been so fruitful, it has come to dominate much of biochemistry.

The molecular interpretation of biological events shows every indication of becoming still more detailed and powerful; therefore the modern student of biochemistry *must* become acquainted with physicochemical theories and methods. These are the subject of this volume. The treatment given here is only a sketch, a preliminary look intended to acquaint the student with a number of ideas and explore a few. It dwells, perhaps too much, on the physicochemical tools the modern biochemist uses. But an understanding of the techniques is fundamental to their intelligent use.

This book presupposes some introduction to physical chemistry. Precisely, I assume that the student has had at least a one-semester course, in which the fundamentals of thermodynamics have been introduced. Having taught such

†*On Growth and Form,* Cambridge University Press, Cambridge, 1917, p. 7.

courses, I am not deluded about the depth of understanding obtained. For this reason, thermodynamics is reviewed in the first chapter.

A word of advice to those who wish to make a career of biochemical research: Go back and learn deeply the fundamentals of mathematics, physics, and chemistry. Then you will be able to show us precisely how the particles of life are "moved, moulded, and conformed."

Chapter One

THERMODYNAMICS AND

BIOCHEMISTRY

Thermodynamics, with its emphasis on heat engines and abstract energy concepts, has often seemed irrelevant to biochemists. Indeed, a conventional introduction to the subject is almost certain to convince the student that much of thermodynamics is sheer sophistry and unrelated to the real business of biochemistry, which is discovering how molecules make organisms work.

But an understanding of some of the ideas of thermodynamics *is* important to biochemistry. In the first place, the very abstractness of the science gives it power in dealing with poorly defined systems. For example, one can use the temperature dependence of the equilibrium constant for a protein denaturation reaction to measure the enthalpy change without knowing what the protein molecule looks like or even its exact composition. And the magnitude and the sign of that change tell us something more about protein molecules. Again, modern biochemists continually use techniques that depend on thermodynamic principles. A scientist may measure the molecular weight of a macromolecule or study its self-association by osmotic pressure measurements. All that is observed is a pressure difference, but the observer *knows* that this difference can be quantitatively interpreted to yield an average molecular weight. To use these physical techniques intelligently, one must understand something of their bases; this is what a good deal of this book is about.

In this first chapter we shall briefly review some of the ideas of thermodynamics that are of importance to biochemistry and molecular biology. While most readers will have taken an undergraduate course in physical chemistry, it has been my experience that this usually leads, insofar as thermodynamics is concerned, to a fairly clear understanding of the first law and some confusion about the second. Since the aim of this section is the use of thermodynamics rather than contemplation of its abstract beauty, we shall emphasize some molecular interpretations of thermodynamic principles. But it should never be forgotten that thermodynamics does not depend, for its rigor, on explicit details of molecular behavior. It is, however, sometimes easier to visualize thermodynamics in this way.

1.1 Heat, Work, and Energy

FIRST LAW OF THERMODYNAMICS

The intention of biochemistry is ultimately to describe certain macroscopic systems, involving multitudes of molecules, in terms of individual molecular properties. The fact is, however, that such systems are so complex that a complete description is beyond the capabilities of present-day physical chemistry. On the other hand, a whole field of study of energy relationships in macroscopic systems has been developed that makes no appeal whatsoever to molecular explanations. This discipline, thermodynamics, allows very powerful and exact conclusions to be drawn about such systems. The laws of thermodynamics are quite exact for systems containing many particles, and this gives a clue as to their origin. They are essentially statistical laws.

The situation is in a sense like that confronting an insurance company; the behavior of the individuals comprising its list of insurees is complex, and an attempt to trace out all of their interactions and predict the fate of any one would be a staggering task. But if the number of individuals is very large, the company can rely with great confidence on statistical laws, which say that so many will perish or become ill in any given period. Similarly, physical scientists can draw from their experiences with macroscopic bodies (large populations of molecules) laws that work very well indeed, even though the laws may leave obscure the mechanism whereby the phenomena are produced. Just as the insurance company with only 10 patrons is in a precarious position (it would not be too unlikely for all 10 people to die next month), so is the chemist who attempts to apply thermodynamics to systems of a few molecules. But 1 mole is 6×10^{23} molecules, a large number indeed. However, a bacterial cell may contain only a small number of some kinds of molecules; this means that some care must be taken when one attempts to apply thermodynamic ideas to systems of this kind [but see Hill (1963)].

For review, let us define a few fundamental quantities.

System. A part of the universe chosen for study. It will have spatial boundaries but may be *open* or *closed* with respect to the transfer of matter. Similarly, it may or may not be thermally insulated from its surroundings. If insulated, it is said to be an *adiabatic* system.

State of the System. The thermodynamic state of a system is clearly defin-able only for systems at equilibrium. In this case, specification of a certain number of variables (two of three variables—temperature, pressure, and volume—plus the masses and identities of all independent chemical substances in the system) will specify the state of the system. In other words, specification of the state is a recipe that allows us to reproduce the system at any time. It is an observed fact that if the state of a system is specified, its properties are given. The properties of a system are of two kinds. *Extensive* properties, such as volume and energy, require for their definition specification of the thermo-dynamic state including the amounts of all substances. *Intensive* properties, such as density or viscosity, are fixed by giving less information; only the *relative* amounts of different substances are needed. For example, the density of a 1 *M* NaCl solution is independent of the size of the sample, though it depends on temperature (T), pressure (P), and the concentration of NaCl.

Thermodynamics is usually concerned with changes between equilibrium states. Such changes may be *reversible* or *irreversible*. If a change is reversible, the path from initial to final state leads through a succession of near-equilibrium states. The system always lies so close to equilibrium that the direction of change can be reversed by an infinitesimal change in the surroundings.

Heat (q). The energy transferred into or out of a system as a consequence of a temperature difference between the system and its surroundings.

Work (w). Any other exchange of energy between a system and its sur-roundings. It may include such cases as volume change against external pressure, changes in surface area against surface tension, electrical work, and so forth.

Internal Energy (E). The energy within the system. In chemistry we usually consider only those kinds of energy that might be modified by chemical pro-cesses. Thus energy involved in holding together the atomic nuclei is generally not counted. The internal energy of a system may then be taken to include the following: translational energy of the molecules, vibrational energy of the molecules, rotational energy of the molecules, the energy involved in chemical bonding, and, the energy involved in nonbonding interactions between mole-cules. Some such interactions are listed in Table 1.1.

TABLE 1.1   *Noncovalent interactions between molecules*

Type of Interaction	Equation[a]	Order of Magnitude[b] (kcal/mole)
Ion–ion	$E = \dfrac{q_1 q_2}{Dr}$	14
Ion–dipole	$E = \dfrac{q_1 \mu_2 \theta}{Dr^2}$	-2 to $+2$
Dipole–dipole	$E = \dfrac{\mu_1 \mu_2 \theta'}{Dr^3}$	-0.5 to $+0.5$
Ion–induced dipole	$E = \dfrac{q_1^2 \alpha_2}{2D^2 r^4}$	0.06
Dispersion[c]	$E = \dfrac{3h\nu_0 \alpha^2}{4r^6}$	0 to 10

[a]In these equations q is charge on an ion, μ is the dipole moment of a dipole, and α is the molecular polarizability (see Chapter 6 for definitions). D is the dielectric constant of the medium and r is the distance between the molecules. The factors θ and θ' are functions of the orientations of dipoles. (See E. A. Moelwyn-Hughes, *Physical Chemistry*, Pergamon Press Ltd., Oxford, 1961, Chap. 7, for details and derivations.)

[b]Calculations were made with the following assumptions. (1) Molecules are 3 Å apart; (2) all charges are 4.8×10^{-10} esu (electron charge); (3) all dipole moments are 2 debye units; and (4) all polarizabilities are 2×10^{-24} cm³. These are typical values for small molecules and ions. The dielectric constant was taken to be 8, a reasonable value for a *molecular* environment; energies would be lower in aqueous solution, where $D \simeq 80$. Since the ion–dipole and dipole–dipole interactions depend strongly on dipole orientation, I have given the *extreme* values. For comparison, covalent bond energies range between 30 and 150 kcal/mole.

[c]Dispersion interactions are between mutually polarizable molecules. The charge fluctuations in the molecules, with frequency ν_0, interact, producing a net interaction. Since this depends so strongly on distance, and becomes important only for very close molecules, a range of values is given.

The internal energy is a function of the state of a system. That is, if the state is specified, the internal energy is fixed at some value regardless of how the system came to be in that state. Since we are usually concerned with energy changes, internal energy is defined with respect to some arbitrarily chosen standard state.

Enthalpy ($H = E + PV$). The internal energy of a system plus the product of its volume and the external pressure exerted on the system. It is also a function of state.

With these definitions, we state the first law of thermodynamics, an expression of the conservation of energy. For a change in state,

$$\Delta E = q - w \tag{1.1}$$

which takes the convention that heat absorbed by a system and work done by a system are positive quantities. For small changes, we write

$$dE = \not{d}q - \not{d}w \qquad (1.2)$$

The slashes through the differential symbols remind us that whereas E is a function of state and dE is independent of the path of the change, q and w do depend on the path. The first law is entirely general and does not depend on assumptions of reversibility and the like.

If the only kind of work done involves change of the volume of the system against an external pressure ($P\,dV$ work),

$$dE = \not{d}q - P\,dV \qquad (1.3)$$

Similarly, we write for the change in the enthalpy of a system the general expression

$$dH = d(E + PV) = dE + P\,dV + V\,dP$$
$$= \not{d}q - \not{d}w + P\,dV + V\,dP \qquad (1.4)$$

For systems doing only $P\,dV$-type work, $\not{d}w = P\,dV$, and

$$dH = \not{d}q + V\,dP \qquad (1.5)$$

Equations (1.3) and (1.5) point up the meaning of dE and dH in terms of measurable quantities. For changes at constant volume

$$dE = dq$$
$$\Delta E = q_v \qquad \text{for a finite change of state} \qquad (1.6)$$

whereas, for processes occurring at constant pressure, (1.5) gives

$$dH = dq$$
$$\Delta H = q_p \qquad \text{for a finite change of state} \qquad (1.7)$$

That is, the heat absorbed by a process at constant volume measures ΔE, and the heat absorbed by a process at constant pressure measures ΔH. These quantities of heat will in general differ, because in a change at constant pressure some energy exchange will be involved in the work done in the change of volume of the system.

The thermochemistry of biological systems is almost always concerned with ΔH, since most natural biochemical processes occur under conditions more nearly approaching constant pressure than constant volume. However, since most such processes occur in liquids or solids rather than in gases, the volume changes are small. To a good approximation, one can often neglect the differences between ΔH and ΔE in biochemistry and simply talk about the "energy change" accompanying a given reaction.

Figure 1.1 summarizes compactly the relationships between the quantities q, w, ΔE, and ΔH. Note that we begin with the perfectly general first law and specialize to particular kinds of processes by adding more and more restrictions.

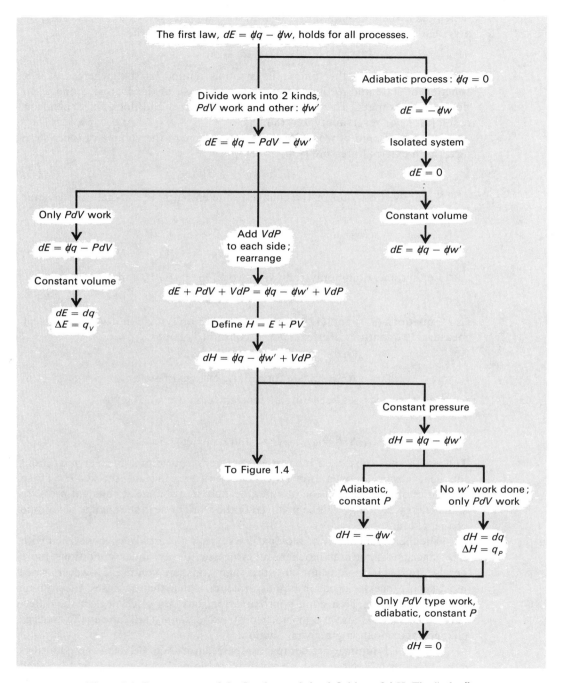

Figure 1.1 Consequences of the first law and the definition of ΔH. The "other" kind of work, dw', may be identified with electrical work, work done in expanding a surface, and so forth.

1.2 Molecular Interpretation of Thermodynamic Quantities

We have seen that from the first law, together with the assertion that the internal energy is a function of the state of a system, powerful and general conclusions can be drawn. Although these have not required molecular models to attest their validity, the student should keep in mind that quantities such as the internal energy and the energy changes in chemical reactions are ultimately expressible in terms of the behavior of atoms and molecules. It will be worth our while to explore the point in more detail.

Suppose that we ask the following question: If we put energy into a system, to give an increase in the internal energy, where has the energy gone? Surely some has appeared as increased kinetic energy, but if the molecules are complex, some must be stored in rotational and vibrational energy and in intermolecular interactions, and perhaps some is accounted for by excited electronic states of a few molecules. Therefore, the question is really one of how the energy is *distributed*.

For thermodynamic properties, we are going to be talking about large numbers of molecules, generally in or near states of equilibrium. The first means that a *statistical* point of view may be taken; we need not follow the behavior of any one molecule. The second implies that we should look for the *most probable* distribution of energy, for we would not expect an equilibrium state to be an improbable one. Although any system might, by momentary fluctuations, occasionally distribute its energy in some improbable way (like having almost all of the energy in a few molecules), the relative occurrence of such extreme fluctuations becomes vanishingly small as the number of molecules becomes very large.

To see the principles involved, let us take a very simple system, a collection of particles that might be thought of, for example, as atoms in a gas or as protein molecules in a solution. Each of these entities is assumed to have a set of energy states available to it, as shown in Figure 1.2. The energy states available to a particle are not to be confused with the thermodynamic "states" of a system of many particles. Rather, they are the quantized states of energy accessible to any particle under the constraints to which the whole collection of particles is subject. Suppose that we have six particles and a total energy of 10ϵ, where ϵ is some unit of energy. Some distributions are shown in Figure 1.2, each of which staisfies the total energy requirement. Now let us say that for any particle, any state is equally probable. This simply means that there is nothing to prejudice a particle to pick a given state. *Then the most probable distribution will be the one that corresponds to the largest number of ways of arranging particles over the states.* If we label the particles, we see that there are only six ways of making state (*a*) and many more ways of making either state (*b*) or (*c*). The number of ways of arranging N particles, n_1, in one group, n_2 in another, and so forth, is

$$W = \frac{N!}{n_1! \, n_2! \, n_3! \ldots n_i! \ldots} \tag{1.8}$$

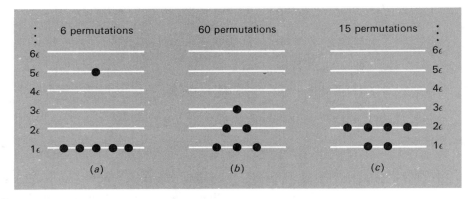

Figure 1.2 Some distributions of particles over energy states, subject to the constraints that $N = 6$ and $E = 10\epsilon$. The numbers are calculated from Equation (1.8) (remember $0! = 1$). The Boltzmann distribution most closely resembles (*b*). Of course, there are not enough particles for it to hold accurately in this simple case.

Since the most probable distribution is the one that corresponds to the largest number of arrangements of particles over energy states, the problem of finding that distribution is a problem of maximizing the number W, subject to the restrictions that N and total energy E are constants. Using standard mathematical techniques for handling such problems, the result for a large number of particles is found to be

$$n_i = n_1 e^{-\alpha(\epsilon_i - \epsilon_1)} \tag{1.9}$$

where n_i is the number of particles in energy state i, n_1 is the number in the lowest state, and ϵ_i and ϵ_1 are the energies of states i and 1, respectively. The constant, a, turns out to be $1/kT$, where k is the Boltzmann constant, the gas constant R divided by Avogadro's number, and T is the absolute temperature. A simple derivation is given in W. J. Moore's *Physical Chemistry* (1972). For a more leisurely, but very clear discussion, see Gurney (1949).

Equation (1.9) is referred to as the *Boltzmann distribution* of energies. It should always be kept in mind that this is not the only possible distribution, and that if we could sample a collection of molecules at any instant we would expect to find deviations from it. It is simply the *most probable* distribution and hence will serve well if the number of particles is large and the system is at, or near, equilibrium.

One more modification of Equation (1.9) should be made. We have written a distribution over energy *states*, whereas a distribution over *levels* would often be more useful. The distinction lies in the fact that levels may be degenerate—there may be several atomic or molecular states corresponding to a given energy level. (The energy levels of the hydrogen atom will serve as one example and the possible different conformational states corresponding to a given energy for a random-coil polymer as another.) If each level contains g_i states (that is, if the degeneracy is some integer g_i), levels should be weighted by this factor. Then

$$n_i = \frac{g_i}{g_1} n_1 e^{-(\epsilon_i - \epsilon_1)/kT} \tag{1.10}$$

where n_i and n_1 now refer to the number of particles in energy *levels* i and 1, respectively.

Equation (1.10) states that if the degeneracies of all states are equal, the lowest states will be the most populated at any temperature. At $T = 0$, $n_i = 0$ for $i > 1$, which means that all particles will be in the lowest level, while as $T \longrightarrow \infty$, the distribution tends to become more and more uniform. At high temperatures, no level is favored over any other, except for the factor of degeneracy. Another useful form of Equation (1.10) involves N, the total number of particles, instead of n_1. If we recognize that $N = \sum_i n_i$ (the sum being taken over all levels), then

$$N = \frac{n_1}{g_1} \sum_i g_i e^{-(\epsilon_i - \epsilon_1)/kT}$$

or

$$\begin{aligned}
\frac{n_i}{N} &= \frac{n_1}{g_1} \frac{g_i e^{-(\epsilon_i - \epsilon_1)/kT}}{(n_1/g_1) \sum_i g_i e^{-(\epsilon_i - \epsilon_1)/kT}} \\
&= \frac{g_i e^{-(\epsilon_i - \epsilon_1)/kT}}{\sum_i g_i e^{-(\epsilon_i - \epsilon_1)/kT}}
\end{aligned} \tag{1.11}$$

The sum in Equation (1.11) is frequently encountered in statistical mechanics. It is sometimes called (for obvious reasons) the *sum over states* and more often (for less obvious reasons) the *molecular partition function*.

Since Equation (1.11) gives the fraction of molecules with energy ϵ_i, it is very useful for calculating average quantities. We shall make use of this idea in subsequent chapters. For example, in Chapter 6 we shall calculate the average orientation of dipolar molecules in an electric field in just this way.

1.3 Entropy, Free Energy, and Equilibrium

SECOND LAW OF THERMODYNAMICS

So far, in discussing chemical and physical processes, we have concentrated on the energetics. We have shown that the first law, a restatement of the conservation of energy, could lead to exceedingly useful and general conclusions about the energy changes that accompany these processes. But one factor has been pointedly omitted—there has been no attempt to predict the *direction* in which changes will occur. Thus the first law allows us to discuss the heat transfer and work accompanying a chemical reaction such as the hydrolysis of adenosine triphosphate (ATP) to yield adenosine diphosphate (ADP):

$$H_2O + ATP \longrightarrow ADP + \text{inorganic phosphate}$$

but gives no indication as to whether or not ATP will spontaneously hydrolyze in aqueous solution. Intuitively, we would expect that under given conditions some particular equilibrium will exist between H_2O, ATP, ADP and inorganic phosphate, but there is no way in which the first law can tell us where that equilibrium lies.

As another example, consider the dialysis experiment shown in Figure 1.3.

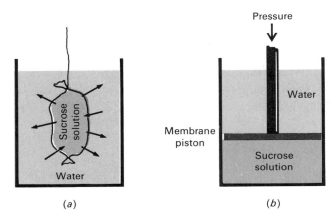

(a)   (b)

Figure 1.3 (*a*) A dialysis experiment in which sucrose will diffuse out of a bag and water into the bag until equilibrium is attained. This process is irreversible. No work is done. (*b*) A way of doing the same experiment reversibly. The membrane piston is impermeable to sucrose and permeable to water. If the pressure on the piston is gradually reduced, the same final state (uniform mixing) approached irreversibily in (*a*) will be approached reversibly. With this arrangement, work will be done.

A solution of sucrose has been placed in a dialysis bag immersed in a container of water. We shall assume that the membrane is permeable to sucrose molecules. Either intuition or a simple experiment will tell us that the system is not at equilibrium. Rather, we know and find that sucrose will diffuse through the membrane until the concentrations inside and outside the bag are equal.

What is it that determines this position of equilibrium and the spontaneous (irreversible) process that leads to it? A first guess might be made from analogy to mechanical processes. A mechanical system reaches a state of equilibrium when the energy of the system is a minimum. (Think of a ball rolling to the bottom of a hill.) But this will clearly not do here. Dilute sucrose solutions are close to ideal (see Chapter 2), which means that the energy of the system is practically independent of concentration. In fact, the interaction of sucrose with water is such that the state of lower concentration is actually of higher energy.

A moment's thought shows that the equilibrium state is favored in this case because of its *higher probability*. If one imagines a vast array of such systems, each of which has been left to itself for a considerable period, it is clear that the great majority will have the sucrose distributed quite uniformly throughout. True, there might be a very, very rare occurrence in which the sucrose concentration was appreciably higher inside or outside. But such systems will be very exceptional, for they would require the remarkable conicidence that a large excess of randomly wandering sucrose molecules happened to be inside or outside of the bag. A closer inspection shows that the main difference between the initial and final states of the system lies in the number of ways (W) in which sucrose molecules can be distributed over the total volume available to them. There are more ways of putting N sucrose molecules into a large volume than into a small volume. Thus one of the determinants of the equilibrium state of a system of many particles is the *randomness* of the system, or the number of ways (W) in which the particles of the system may be distributed, whether it be over levels of energy or, as in this instance, over the volume of the system.

For this reason, we define a function of state, which we shall call the *entropy:*

$$S = k \ln W \qquad (1.12)$$

where k is the Boltzmann constant. The logarithm of W is chosen for the following reason: We wish the entropy to be an extensive property; that is, we wish the S of a system that is made up of two parts (1 and 2) to be $S_1 + S_2$. Now if W_1 is the number of ways of distributing the particles in part 1 in the particular state and W_2 the corresponding quantity for part 2, the number of ways for the whole system in this state is $W_1 W_2$:

$$S = k \ln W = k \ln W_1 W_2 = k \ln W_1 + k \ln W_2 = S_1 + S_2 \qquad (1.13)$$

This description of S, although exceedingly useful for certain problems, is not clearly related to measurable physical quantities. This may be accomplished in the following way: Let us assume that the molecules in a system have available a certain set of energy states and are distributed over these states according to the Boltzmann distribution

$$n_i = n_1 e^{-(\epsilon_i - \epsilon_1)/kT} \qquad (1.14)$$

Now

$$W = \frac{N!}{n_1! n_2! n_3! \ldots n_n!} \qquad (1.15)$$

$$\ln W = \ln(N!) - \ln(n_1!) - \ln(n_2!) - \cdots \qquad (1.16)$$

or, using Stirling's approximation ($\ln n! \simeq n \ln n - n$),

$$\ln W \simeq N \ln N - N - n_1 \ln n_1 + n_1 - n_2 \ln n_2 + n_2 - \cdots \qquad (1.17)$$

Since $\sum_i n_i = N$, we obtain

$$\ln W = K - \sum_i n_i \ln n_i \qquad (1.18)$$

where K is the constant $N \ln N$.

Let us now consider an infinitesimal change in the state of the system. This is assumed to be a *reversible* change; that is, the system does not depart appreciably from a state of equilibrium but simply changes from the Boltzmann distribution appropriate to the initial state to that appropriate to the infinitesimally different new state:

$$dS = kd \ln W \tag{1.19}$$

$$= -k \, d \sum_i n_i \ln n_i = -k \sum_i n_i \, d \ln n_i - k \sum_i \ln n_i \, dn_i \tag{1.20}$$

$$= -k \sum_i \ln n_i \, dn_i \tag{1.21}$$

The first term in Equation (1.20) vanishes because it equals $\sum_i dn_i = d \sum_i n_i$ and the total number of particles is constant. The expression for dS can then be evaluated by remembering that $n_i = n_1 e^{-(\epsilon_i - \epsilon_1)/kT}$, so that

$$\ln n_i = \ln n_1 - \frac{\Delta \epsilon_i}{kT} \tag{1.22}$$

where $\Delta \epsilon_i$ is used to abbreviate $(\epsilon_i - \epsilon_1)$. Again dropping a term that is the sum of dn_i, we obtain

$$dS = k \left(\frac{\sum \Delta \epsilon_i}{kT} \right) dn_i = \frac{1}{T} \sum_i \Delta \epsilon_i \, dn_i \tag{1.23}$$

Now the sum in Equation (1.23) is simply the heat absorbed in the process, for if a system has a given set of *fixed* energy levels and a reversible change in the population of the levels occurs, this must involve the absorption or release of heat from the system:

$$dS = \frac{dq_{rev}}{T} \tag{1.24}$$

This is the classic definition of an entropy change. For a finite isothermal process, we say that

$$\Delta S = \frac{q_{rev}}{T} \tag{1.25}$$

The heat, q_{rev}, is a perfectly defined quantity, since the change is required to be reversible.

Equation (1.25) states that if we wish to calculate the entropy change in the transition between state 1 and state 2 of a system, it is only necessary to consider a reversible path between the two states and calculate the heat absorbed or evolved. As a simple example, we may consider the case of an isothermal, reversible expansion of 1 mole of an ideal gas from volume V_1 to volume V_2. For each infinitesimal part of the process, $dE = 0$ (since the gas is ideal and its energy is independent of its volume). Then

$$dq = dw = P \, dV = \frac{RT \, dV}{V} \tag{1.26}$$

and

$$\frac{q_{rev}}{T} = \frac{RT}{T} \int_{V_1}^{V_2} \frac{dV}{V} = R \ln \left(\frac{V_2}{V_1} \right) \tag{1.27}$$

so

$$\Delta S = R \ln \left(\frac{V_2}{V_1} \right) \tag{1.28}$$

The entropy of the gas increases in such an expansion.

We can arrive at exactly the same result from the statistical definition of entropy. Suppose that we divide the volume V_1 into n_1 cells, each of volume V: therefore $V_1 = n_1 V$. The larger volume V_2 is divided into n_2 cells of the same size: $V_2 = n_2 V$. Putting one molecule into the initial system of volume V_1, there are n_1 ways in which this can be done. For two molecules there will be n_1^2 ways. For an Avogadro's number ($\mathfrak{N}$) of molecules, the number of ways to put them into V_1 is $W = n_1^{\mathfrak{N}}$. If the larger volume V_2, containing V_2/V cells, is occupied by the same $\mathfrak{N}$ molecules, $W_2 = n_2^{\mathfrak{N}}$. From Equation (1.12),

$$\Delta S = k \ln n_2^{\mathfrak{N}} - k \ln n_1^{\mathfrak{N}}$$

$$= k \ln \left(\frac{n_2}{n_1} \right)^{\mathfrak{N}} = R \ln \left(\frac{n_2}{n_1} \right) = R \ln \left(\frac{V_2/V}{V_1/V} \right) \tag{1.29}$$

where we have made use of the fact that $\mathfrak{N}k = R$. Then

$$\Delta S = R \ln \left(\frac{V_2}{V_1} \right) \tag{1.30}$$

Note that the cell volume V has canceled out of the final result; it is just a device to allow us to compare the number of ways of making the systems 1 and 2. Note also that it was necessary for the constant in Equation (1.12) to have the value k (Boltzmann's constant) in order that the two methods of approach would lead to the same numerical value of ΔS.

A very similar calculation could be carried out for our example of sucrose diffusing out of a dialysis bag. In this case, however, the calculation from q_{rev}/T would be more difficult, for we would have to imagine some reversible way in which to carry out the process. One such way is depicted in Figure 1.3(*b*). The dialysis bag is replaced by a membrane piston, permeable to water, but not to sucrose. To the piston is applied a pressure equal to the osmotic pressure of the solution (see Chapter 2). If this pressure is gradually reduced, the piston will rise, doing work on the surroundings and diluting the sucrose solution. Since once again the ideality of the system requires $\Delta E = 0$, heat must be absorbed to keep the system isothermal.

Both the classic and statistical calculation will again lead to the same result, which is formally identical to that for the gas expansion. Per mole, we have

$$\Delta S = R \ln \left(\frac{V_2}{V_1} \right) = R \ln \left(\frac{C_1}{C_2} \right) \tag{1.31}$$

where C_1 and C_2 are concentrations.

A very similar calculation allows us to determine the entropy of mixing for an ideal solution made up of N components. Assume that there are N_i molecules of each component, making a total number of N_0 molecules in the whole solution: $N_0 = \sum_{i=1}^{N} N_i$. We assume that the molecules are about the same size, so that each can occupy a "cell" of the same volume. Since the mixture is assumed to be ideal, there will be no volume change in mixing, and the total number of cells in the solution will be N_0, the total number of molecules. The entropy of mixing is defined as

$$\Delta S_m = S \text{ (solution)} - S \text{ (pure components)} \qquad (1.32)$$

We are concerned only with mixing entropy, so we may say that each of these quantities is given by $k \ln W$, where W is the number of distinguishable ways of putting molecules in cells. So

$$\Delta S_m = k \ln \frac{W \text{ (solution)}}{W \text{ (pure components)}} \qquad (1.33)$$

But all arrangements of molecules in the pure components are the same, so W (pure components) $= 1$, and $\Delta S_m = k \ln W$ (solution). But W (solution) is just the number of ways of arranging N_0 cells into groups with N_1 of one type, N_2 of another, and so forth. Therefore, we have the familiar expression

$$\Delta S_m = k \ln \frac{N_0!}{N_1! N_2! \cdots N_N!} \qquad (1.34)$$

Since all of the N's are large, we may use Stirling's approximation:

$$\Delta S_m \simeq k(N_0 \ln N_0 - N_0 - N_1 \ln N_1 \\ + N_1 - N_2 \ln N_2 + N_2 - \cdots) \qquad (1.35)$$

or, since $\sum_i N_i = N_0$,

$$\Delta S_m = k(N_0 \ln N_0 - \sum_i N_i \ln N_i) \qquad (1.36)$$

or

$$\Delta S_m = k(\sum_i N_i \ln N_0 - \sum_i N_i \ln N_i) \qquad (1.37)$$

$$= -k \sum_i N_i \ln \frac{N_i}{N_0} = -k \sum_i N_i \ln X_i \qquad (1.38)$$

where X_i is the mole fraction of i. Multiplying and dividing by Avogadro's number, we obtain the final result:

$$\Delta S_m = -R \sum_i n_i \ln X_i \qquad (1.39)$$

where n_i is the number of moles of component i. This expression says that even if there is no interaction between the molecules, the entropy of a mixture is always greater than that of the pure components, since all X_i's are less than unity, and their logarithms will be negative quantities.

It is evident from the above examples and from Equation (1.12) that the entropy can be considered as a measure of the randomness of a system. A crystal is a very regular and nonrandom structure; the liquid to which it may be melted is much more random and has a higher entropy. The entropy change in melting can be calculated as

$$\Delta S = \frac{q_{rev}}{T} = \frac{\Delta H_{melting}}{T_{melting}} \tag{1.40}$$

Since the melting of a crystalline solid will always be an endothermic process and since T is always positive, the entropy will always increase when a solid is melted. Similarly, a *native* protein molecule is a highly organized, regular structure. When it is *denatured*, an unfolding and unraveling takes place; this corresponds to an increase in entropy. Of course, this is not the whole story; upon denaturation the interaction of solvent with the protein may change in such a way either to add or subtract from this entropy change.

Whenever a substance is heated, the entropy will increase. We can see this in the following way: At low temperatures, only the few lowest of the energy levels available to the molecules are occupied; there are not too many ways to do this. As T increases, more levels become available and the randomness of the system increases. The entropy change can, of course, be calculated by assuring that the heating is done in a reversible fashion. Then, if the process is at constant pressure, $dq_{rev} = C_p\, dT$, where C_p is the constant pressure heat capacity

$$dS = \frac{C_p\, dT}{T} \tag{1.41}$$

or, in heating a substance from temperature T_1 to T_2,

$$\Delta S = \int_{T_1}^{T_2} \frac{C_p}{T}\, dT \tag{1.42}$$

If C_p is constant,

$$\Delta S = C_p \ln\left(\frac{T_2}{T_1}\right) \tag{1.43}$$

The definition of the entropy change in an infinitesimal, reversible process as dq_{rev}/T allows us a reformulation of the first law for reversible processes, since

$$dE = dq - dw = T\, dS - P\, dV \tag{1.44}$$

This leads to one precise answer to a question posed at the beginning of this section: What are the criteria for a system to be at equilibrium? If, and only if, a system is at equilibrium, an infinitesimal change will be reversible. We may write from Equation (1.44) that if the volume of the system and its energy are held constant,

$$dS = 0 \tag{1.45}$$

for a reversible change. This means that S must be either at a maximum or minimum for a system at constant E and V to be at equilibrium. Closer analysis reveals that it is the former. *If we isolate a system* (keep E and V constant), then it will be at equilibrium only when the entropy reaches a maximum. In any nonequilibrium condition the entropy will be spontaneously increasing toward this maximum. This is equivalent to the statement that an isolated system will approach a state of maximum randomness. This is the most common way of stating the *second law of thermodynamics.*

Isolated systems kept at constant volume are not of great interest to biochemists. Most of our experiments are carried out under conditions of constant temperature and pressure. To elucidate the requirements for equilibrium in such circumstances, a new energy function must be defined which is an explicit function of T and P. This is the *Gibbs free energy, G*:

$$G = H - TS \tag{1.46}$$

We obtain the general expression for a change dG by differentiating Equation (1.46):

$$dG = dH - T\,dS - S\,dT$$
$$= dE + V\,dP + P\,dV - S\,dT - T\,dS \tag{1.47}$$

This collection of terms simplifies if we note that $dE = T\,dS - P\,dV$ if the process causing dG is reversible. Then

$$dG = V\,dP - S\,dT \qquad \text{reversible process} \tag{1.48}$$

And now, if the system is also constrained to constant T and P,

$$dG = 0 \qquad \text{reversible process} \tag{1.49}$$

This may be interpreted in the same way as Equation (1.45). The value of G is at an extremum when a system constrained to constant T and P reaches equilibrium. In this case it will be a minimum.

A similar quantity, the *Helmholtz free energy*, is defined as $A = E - TS$. It is of less use in biochemistry, because $dA = 0$ for the less frequently encountered conditions of constant T and V. By *free energy* we shall mean the Gibbs free energy. Some of these consequences of the second law are worked out in compact form in Figure 1.4.

The Gibbs free energy is of enormous importance in deciding the direction of processes and positions of equilibrium in biochemical systems. If the free-energy change calculated for a process under particular conditions is found to be negative, that process is spontaneous, for it leads in the direction of equilibrium. Furthermore, ΔG, as a combination of ΔH and ΔS, emphasizes the fact that both energy minimization and entropy (randomness) maximization play a part in determining the position of equilibrium. As an example, consider the denaturation of a protein or polypeptide:

$$\Delta G_{\text{den}} = \Delta H_{\text{den}} - T\,\Delta S_{\text{den}} \tag{1.50}$$

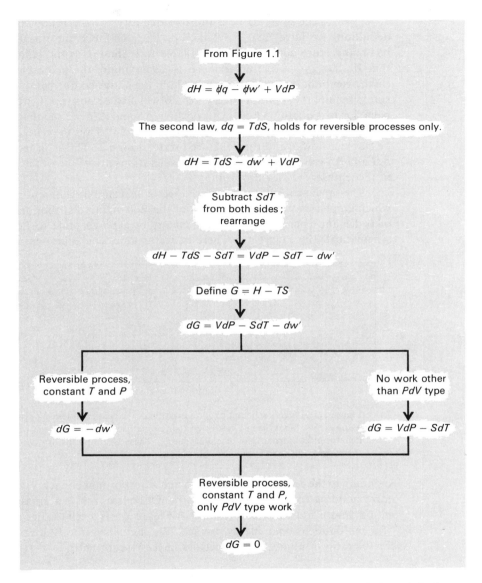

Figure 1.4 A continuation of Figure 1.1 to summarize some relations derived from the second law. Note that here q, w, and w' are those for reversible processes only.

Now we expect ΔS_{den} to be a positive quantity for the following reason: By definition, we have $\Delta S_{den} = R \ln (W_{den}/W_{native})$. Since the denatured protein has many more conformations available to it than does the native, it follows that $W_{den}/W_{native} \gg 1$. Furthermore, the breaking of the favorable interactions (hydrogen bonding and the like) that hold the native conformation together will surely require the input of energy, so ΔH will also be positive. If we examine the behavior of Equation (1.50) with positive ΔS and ΔH, we see that at some low T, $\Delta G > 0$, whereas at sufficiently high T, $\Delta G < 0$. Thus the native state should be stable at low T and the denatured state at high T. This is in fact found, and ΔH and ΔS estimated from simple considerations are of the order of magnitude of those observed (see Problem 5).

The same problem can be considered from a slightly different point of view, which emphasizes the close relation of the Boltzmann distribution to problems of order and energy. Consider the energy of a polypeptide to be represented schematically by Figure 1.5. There is, we assume, one energy minimum corre-

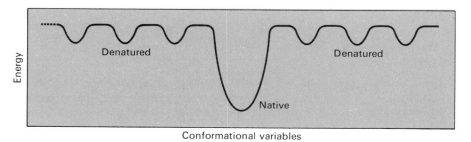

Conformational variables

Figure 1.5 A *highly* schematic representation of the energy states of a protein. We assume that the molecule has only a single native state and a large number of denatured conformations of equal energy. These assumptions are certainly far too severe, but they allow a simple calculation.

sponding to the ordered conformation and a host of higher energy conformations corresponding to the random-coil form. Of course, this is a gross oversimplification, and it must be understood that Figure 1.5 is a schematic representation of a multidimensional energy surface. We may write the Boltzmann equation for the ratio of numbers of molecules in the two forms as

$$\frac{n_{den}}{n_{native}} = \frac{g_d}{g_n} e^{-(\epsilon_d - \epsilon_n)/kT} \tag{1.51}$$

We can then identify the degeneracies of these levels with the number of conformations corresponding to each:

$$g_n = W_n = 1 \tag{1.52}$$

$$g_d = W_d \gg 1 \tag{1.53}$$

Now, we can rewrite Equation (1.51) as

$$\frac{n_d}{n_n} = e^{\ln (W_d/W_n)} e^{-(\epsilon_d - \epsilon_n)/kT}$$

$$= e^{-[(\epsilon_d - \epsilon_n) - kT \ln (W_d/W_n)]/kT} \qquad (1.54)$$

or, putting quantities on a molar basis,

$$\frac{n_d}{n_n} = e^{-[(E_d - E_n) - RT \ln (W_d/W_n)]/RT} \qquad (1.55)$$

Both sides of this equation are recognizable. In the exponent, if we neglect the difference between ΔE and ΔH and note that the quantity $R \ln (W_d/W_n)$ is an entropy change, we have

$$\frac{n_d}{n_n} = e^{-(\Delta H - T \Delta S)/RT}$$

$$= e^{-\Delta G/RT} \qquad (1.56)$$

The left side is simply the equilibrium constant for the reaction. Therefore,

$$\ln K = \ln \left(\frac{n_d}{n_n}\right) = -\frac{\Delta G}{RT} \qquad (1.57)$$

This equation will be familiar to most readers, and we shall derive it more rigorously in Chapter 3. Here, it points out again the role of randomness: The denatured state may be favored, in some circumstances, just because it corresponds to more states of the molecules. Conversely, it should be noted that only because of the energetically favorable residue interactions does the native form described above exhibit stability. Were these not present, the entropy change would drive the reactions toward the random coil at *any* temperature above absolute zero.

It should not be supposed that the stability of real macromolecular conformations is as simple as implied above. For one thing, the solvent cannot, in general, be neglected, and solute–solvent interactions may play a very important role. To take a particular example, it has been argued that the stability of some macromolecules may derive in part from *hydrophobic bonding*. Such bonding may lead to a very different temperature dependence of stability. In the breaking of a hydrophobic bond, nonpolar groups are separated from one another and put in contact with the solvent. In aqueous solution, such groups are expected to become surrounded by shells of "icelike" water. The immobilization of a large number of water molecules should correspond to an entropy decrease—perhaps so much as to override the entropy increase accompanying the macromolecule's gain of conformational freedom. If this is the case, the overall entropy change for the transition from the ordered conformation of the macromolecule to the random one would involve a *decrease* in entropy. Then the temperature dependence should be a reverse of the above example; *low* temperatures should favor disorganization. Such behavior may be being observed when we see that some multisubunit proteins dissociate into their individual subunits at temperatures around 0°C.

PROBLEMS†

[*Note:* The problems in this chapter are intended mainly for review of the student's previous experience with physical chemistry. They may draw on details that have not been mentioned in this chapter.]

1. Calculate q, w, ΔE, and ΔH for the isothermal expansion of 2 moles of an ideal gas from 16 liters to 95 liters at 0°C.

2. The standard-state enthalpy change for the oxidation of palmitic acid

$$CH_3(CH_2)_{14}COOH(s) + 24O_2(g) \longrightarrow 17CO_2(g) + 16H_2O(l)$$

is -2380 kcal.
    (*a*)   Calculate ΔE.
    (*b*)   Calculate the work done when 1 mole is oxidized at 1 atm pressure.

3. The heat of melting of ice at 1 atm and 0°C is $+1.4363$ kcal/mole. The density of ice under these conditions is 0.917 g/cm³ and the density of water 0.9998 g/cm³.
    (*a*)   Calculate $\Delta H - \Delta E$.
    (*b*)   Calculate ΔS.
  (c)   What will the melting point be at 10 atm? Derive necessary equations, make necessary assumptions.

4. Assuming (see Chapter 2) that the osmotic pressure is given by $\pi = RTC/M$, work out the ΔS for the process depicted in Figure 1.3. Assume that you start with 1 g of sucrose in 100 ml and increase the volume to 1 liter.

*5. Assuming that a polypeptide chain has only one α-helical conformation and that there are three possible orientations for each amino acid residue in the random-coil state, calculate ΔS for the conformational change

$$\alpha \text{ helix} \rightleftharpoons \text{random coil}$$

for a polypeptide of 100 residues. What value of ΔH per residue would be required to make the melting point (the temperature at which the equilibrium constant equals 1) be 50°C? Compare with the hydrogen-bond energy, estimated to be 0 to 3 kcal/mole.

6. Two energy levels of a molecule are separated by 1×10^{-15} erg. The degeneracy of the higher level is twice that of the lower. Calculate
    (*a*)   The relative populations of these levels at 0°C.
    (*b*)   The temperature at which they will be equally populated.

*7. Closed circular DNA molecules can be supercoiled. The energy required to put i supercoil twists into a mole of DNA molecules is given approximately by $\Delta E = Ki^2$, where $K \simeq 0.1$ kcal/mole. Consider a circular DNA molecule which has been broken, and the ends rejoined. Show that the number of molecules with i supercoil twists after rejoining is given by a Gaussian distribution about $i = 0$, and calculate the relative numbers of molecules with 0, 1, 2, and 10 supercoil twists, at 37°C.

†In this and subsequent chapters the problems judged to be more difficult are marked with an asterisk. Note that answers for all odd-numbered problems are given in the back of the book.

REFERENCES

Thermodynamics and elementary statistical mechanics are well treated in any number of the better physical chemistry texts. I recommend especially:

Moore, W. J.: *Physical Chemistry*, 4th ed., Prentice-Hall, Inc., Englewood Cliffs, N.J., 1972. More detailed and rigorous than most of the available texts at this level.

There are also several texts directed toward students of the biological sciences. I have found the following to be especially good:

Eisenberg, D. and D. Crothers: *Physical Chemistry with Applications to the Life Sciences*, The Benjamin-Cummings Publishing Co., Menlo Park, Calif., 1979.

Tinoco, I., Jr., K. Sauer, and J. C. Wang: *Physical Chemistry: Principles and Applications in Biological Sciences*, Prentice-Hall, Inc., Englewood Cliffs, N.J., 1978.

In the realm of more specialized books, the following have been especially useful:

Denbigh, K. G.: *The Principles of Chemical Equilibrium*, Cambridge University Press, Cambridge, 1955. A very different book—discursive, somewhat long, but beautifully clear.

Gibbs, J. W.: *The Collected Works of J. W. Gibbs*, Yale University Press, New Haven, Conn., 1948. Not a useful text but worth examining for the elegance of the analysis.

Gurney, R. W.: *Introduction to Statistical Mechanics*, McGraw-Hill Book Company, New York, 1949. An excellent book for the beginner in statistical thermodynamics.

Hill, T. L.: *Thermodynamics of Small Systems*, W. A. Benjamin, Inc., New York, 1963. The application of statistical methods to a number of kinds of problems of interest to the biochemist.

Kirkwood, J. G. and I. Oppenheim: *Chemical Thermodynamics*, McGraw-Hill Book Company, New York, 1961. A high-level, authoritative treatment; terse.

Klotz, I.: *Energetics in Biochemical Reactions*, Academic Press, Inc., New York, 1957. A very brief, nonmathematical introduction to thermodynamics for biochemists. Some nice examples and explanations.

Morowitz, H. J.: *Entropy for Biologists*, Academic Press, Inc., New York, 1970. Actually a brief (and good) thermodynamics text.

Chapter Two

SOLUTIONS OF

MACROMOLECULES

The development of a true molecular biology was long inhibited by fundamental misunderstandings about the nature of solutions of proteins, carbohydrates, and other large molecules. In the early 1900s, recognition of some of the remarkable features of the behavior of such solutions led to the idea that these were not true solutions at all. Since they exhibited no measurable freezing-point depression and since the solutes could not easily be caused to crystallize and exhibited very slow diffusion, such solutions were called *colloids*. This carried the unfortunate connotation that the well-known physicochemical laws for "true" solutions should not apply and that the application of conventional solution thermo-dynamics in this field was of doubtful validity.

Of course, we now know that proteins, and high polymers in general, form true solutions. The freezing-point depression is indeed small and diffusion is slow, but these are only consequences of the high molecular weights of the solutes. Crystallization is more difficult but possible in almost every case. There is only one cause for reservation, and that is easily disposed of. Thermodynamic descriptions of solutions are dependent on there being a very large number of solute particles in the sample observed, so that macroscopic fluctuations in properties will be very unlikely. Is this condition satisfied? If we consider an extreme case, a solution containing 0.01 mg/ml of a virus of molecular weight 100 million, we shall still find roughly 10^{10} particles per milliliter. This is a number large enough that we need not worry about fluctuations. It is only when we begin considering volumes comparable to that of a single cell that such questions need be raised.

With this preface, we shall first review the ways in which the thermodynamics of solutions is described and then apply these concepts to the description of solutions of macromolecules.

2.1 Some Fundamentals of Solution Thermodynamics

A solution is a single-phase system containing more than one component. A *component* is an independently variable chemical substance. It should be noted that this definition of component is strict and that a solution will frequently contain fewer components than molecular species that might be present. This will occur whenever chemical equilibria exist in the solution. To take an example, consider a solution containing water, the protein hemoglobin (Hb), and dissolved oxygen. Since each hemoglobin molecule can bind one, two, three, or four oxygen molecules, there will be a number of molecular species potentially or actually present: H_2O, Hb, O_2, HbO_2, $Hb(O_2)_2$, $Hb(O_2)_3$, and $Hn(O_2)_4$. Yet if the binding reactions are in equilibrium, there are only three *independently variable* substances, or components. Specification of the solvent plus any two of the others would, via the equilibrium relationships, specify the rest. Thus the solution contains three and only three components. Specification of the amounts of three of the substances, together with the temperature and pressure, will completely define the state of the system. It is important to emphasize that such simplification is possible only if the system is in equilibrium. If, in the case above, the oxygenation or deoxygenation reactions were very slow, we might perturb the system (put in some more oxygen, for example) and observe the system in a nonequilibrium state. In this case, more than three components would be needed to describe the system.

The description of the state of a solution by stipulation of T, P, and the amounts of n components may be thought of as a recipe. It means that if, on two or more occasions, we fix these variables at the same values, we shall find exactly the same properties, both extensive and intensive. We may logically ask, then, how some extensive property, such as the volume, will depend on the amounts of the various components. It would be extremely naive to assume simple additivity. That is, we would not expect that n_i moles of each component of molar volumes V_i would, when mixed, give a total volume $V = \sum_i n_i V_i$. In many cases volume changes on mixing can easily be observed. The true situation is easy to see; if we add a small amount of component i to a mixture, the change produced in an extensive property will depend not only on the amount and nature of the substance added but also on the composition of the mixture to which addition is made. This leads to the definition of *partial molar* and *partial specific* quantities.

Considering any extensive property, X, we define the partial molar quantity, $\bar{X}_i$, as

$$\bar{X}_i = \left(\frac{\partial X}{\partial n_i}\right)_{T,P,n_{j\neq i}} \tag{2.1}$$

where n_i is the number of moles of component i. The subscripts indicate that we hold T, P, and the amounts of all components except i constant. The partial specific quantity, $\bar{x}_i$, is given by

$$\bar{x}_i = \left(\frac{\partial X}{\partial g_i}\right)_{T,P,g_{j\neq i}} \tag{2.2}$$

where g_i represents grams of component i. The latter quantities are frequently used in macromolecular chemistry, since we often do not know the molecular weight and hence the number of moles in a sample.

To make these definitions concrete, let us consider the volume of a solution, V, as the extensive property. The partial molar volume ($\bar{V}_i$) and partial specific volume ($\bar{v}_i$) of component i are then given by

$$\bar{V}_i = \left(\frac{\partial V}{\partial n_i}\right)_{T,P,n_{j\neq i}} \tag{2.3}$$

and

$$\bar{v}_i = \left(\frac{\partial V}{\partial g_i}\right)_{T,P,g_{j\neq i}} = \frac{\bar{V}_i}{M_i} \tag{2.4}$$

respectively.

These partial quantities are intensive variables, independent of the size of the system. Thus $\bar{V}_i$ will depend on the composition of the mixture of which i is a part and on T and P but not on the amount of the solution. This leads to a very important result, which bears on the original question as to how *extensive* properties may be calculated. Suppose that we imagine putting together a volume V of solution by starting from nothing and adding infinitesimal increments of the n components, always in the proportions to be found in the final mixture. That is, we make $dg_1 = g_1' \, d\lambda$, $dg_2 = g_2' \, d\lambda$, $\ldots$, $dg_i = g_i' \, d\lambda$, where g_1', g_2', $\ldots$, g_i' are constants equal to the number of grams of each substance in the final mixture and λ is a dummy variable going from 0 to 1. With this method of assembly, the composition will always be the same, and the $\bar{v}$'s will be constant during construction of the solution. Now

$$dV = \bar{v}_1 dg_1 + \bar{v}_2 \, dg_2 + \cdots + \bar{v}_i \, dg_i + \cdots + \bar{v}_n \, dg_n$$
$$= g_1'\bar{v}_1 \, d\lambda + g_2'\bar{v}_2 \, d\lambda + \cdots + g_i'\bar{v}_i \, d\lambda + \cdots + g_n'\bar{v}_n \, d\lambda \tag{2.5}$$

Since both the g_i''s and the $\bar{v}_i$'s are constants, we may integrate easily:

$$V = (g_1'\bar{v}_1 + g_2'\bar{v}_2 + \cdots + g_i'\bar{v}_i + \cdots + g_n'\bar{v}_n) \int_0^1 d\lambda \tag{2.6}$$

$$= \sum_{i=1}^{n} g_i'\bar{v}_i \tag{2.7}$$

The total extensive quantity is the sum of the partial specific quantities each multiplied by the number of grams of the appropriate component. Similarly, with partial molar quantities

$$X = \sum_{i=1}^{n} n_i \bar{X}_i \tag{2.8}$$

These are the appropriate summation rules for real solutions.

A quantity that will be of the utmost significance in many of our problems is the partial molar free energy. We shall again and again be concerned with equilibria of multicomponent systems at constant T and P. Hence the free energy of the solution will be the property of interest, and we must know the contribution of each component to the total free energy. The partial molar free energy is given the special symbol μ and is defined as

$$\bar{G}_i = \mu_i = \left(\frac{\partial G}{\partial n_i}\right)_{T,P,n_{j \neq i}} \tag{2.9}$$

The total free energy of the solution is then

$$G = \sum_{i=1}^{n} n_i \mu_i \tag{2.10}$$

The partial molar free energy is often called the *chemical potential*. This term is appropriate, for as we shall see, differences in μ may be regarded as the driving forces for such processes as chemical reactions or diffusion in which changes in the amounts of chemical substances occur.

The change in G for an infinitesimal, reversible change in the state of an open system (in which amounts of components can change) is given by the general formula

$$dG = \left(\frac{\partial G}{\partial T}\right)_{P,n_i} dT + \left(\frac{\partial G}{\partial P}\right)_{T,n_i} dP + \sum_i \left(\frac{\partial G}{\partial n_i}\right)_{T,P,n_j} dn_i \tag{2.11}$$

or

$$dG = -S \, dT + V \, dP + \sum_i \mu_i \, dn_i \tag{2.12}$$

or, at constant T and P,

$$dG = \sum_i \mu_i \, dn_i \tag{2.13}$$

From the results above we can obtain an additional important and useful theorem, the Gibbs–Duhem equation. If we differentiate Equation (2.10) generally,

$$dG = \sum_i \mu_i \, dn_i + \sum_i n_i \, d\mu_i \tag{2.14}$$

Combining with Equation (2.13), we have, at constant T and P,

$$\sum_i n_i \, d\mu_i = 0 \tag{2.15}$$

This means that variations of μ_i are not independent of one another and that we can express the chemical potential of one component in terms of those of the others. For example, in a two-component solution we can write an equation for the chemical potential of the solute in terms of that of the solvent, or vice versa.

The importance of the chemical potential arises from the fact that it measures that increment of free energy accompanying an infinitesimal change in the amount of one particular component in a system. Thus it is immediately applicable to discussions of phase equilibria and chemical equilibria. Two very general statements, on which most of our use of the chemical potential will be based, are given below:

1. If a number of phases are in equilibrium, the chemical potential of a given component will have the same value in all phases in which that component is present. While the general proof will not be given here, the principle involved is obvious: Transfer of an infinitesimal amount of a component between two phases (α and β) in equilibrium at constant T and P must involve a zero free-energy change; this can only be so if μ_i is the same in both phases, since, from Equation (2.13),

$$dG = \mu_i^\alpha \, dn_i^\alpha + \mu_i^\beta \, dn_i^\beta \tag{2.16}$$

but since

$$dn_i^\alpha = -dn_i^\beta \quad \text{and} \quad dG = 0$$

we have

$$0 = (\mu_i^\alpha - \mu_i^\beta) \, dn_i^\alpha \tag{2.17}$$

$$\mu_i^\alpha = \mu_i^\beta \tag{2.18}$$

This principle will be the starting point for discussion of such problems as membrane equilibria.

2. In the general chemical equilibrium,

$$a\mathrm{A} + b\mathrm{B} + \cdots \rightleftharpoons g\mathrm{G} + h\mathrm{H} + \cdots \tag{2.19}$$

where a, b, and so forth, are stochiometric numbers, the chemical potentials of the components A, B, and so forth, must obey the relationship

$$a\mu_\mathrm{A} + b\mu_\mathrm{B} + \cdots = g\mu_\mathrm{G} + h\mu_\mathrm{H} + \cdots \tag{2.20}$$

if the system is to be at chemical equilibrium.

This result can be demonstrated easily. Suppose that the system is kept at constant T and P but that an infinitesimal displacement from equilibrium occurs. Since the free energy must have been at a minimum if the system was initially at equilibrium, $dG = 0$. Furthermore, according to the stochiometric equation (2.19), the dn_i's are not independent but must be in the proportions

$$\frac{dn_\mathrm{A}}{a} = \frac{dn_\mathrm{B}}{b} = \cdots = -\frac{dn_\mathrm{G}}{g} = -\frac{dn_\mathrm{H}}{h} = \cdots \tag{2.21}$$

Now dG may be written as

$$dG = \mu_\mathrm{A} \, dn_\mathrm{A} + \mu_\mathrm{B} \, dn_\mathrm{B} + \cdots + \mu_\mathrm{G} \, dn_\mathrm{G} + \mu_\mathrm{H} \, dn_\mathrm{H} + \cdots \tag{2.22}$$

which gives

$$0 = a\mu_A \frac{dn_A}{a} + b\mu_B \frac{dn_B}{b} + \cdots + g\mu_G \frac{dn_G}{g} + h\mu_H \frac{dn_H}{h} + \cdots \qquad (2.23)$$

or from Equation (2.21)

$$a\mu_A + b\mu_B + \cdots = g\mu_G + h\mu_H + \cdots \qquad (2.24)$$

These statements about the chemical potential are of importance because the chemical potential of a substance in a mixture depends on its concentration. It is this dependence that allows us to link the very general statements concerning equilibrium to specific experimental problems, where we observe variations in the concentrations of substances. The question then is: How does μ_i depend on the concentration of i? To answer this, we must inquire a bit more into the nature of solutions.

In physical chemistry, the distinction is made between *ideal* and *nonideal* solutions. This is useful, since ideal behavior leads to exceedingly simple laws, which serve as prototypes for the more complex laws describing real, nonideal solutions. The physical chemist customarily defines an ideal solution as one in which all components follow Raoult's law; that is, the vapor pressure of each component is strictly proportional to the mole fraction of that component over the entire concentration range. We shall adopt a somewhat different definition, which leads to the same result but enables us to see directly how the chemical potential depends on concentration. An ideal solution will be defined as one in which:

1. The enthalpy change in mixing is zero.

$$\Delta H_m = 0 \qquad (2.25)$$

2. The entropy change in mixing is given by the simple statistical rule (see Chapter 1)

$$\Delta S_m = -R \sum_{i=1}^{i} n_i \ln X_i \qquad (2.26)$$

where the X_i's are mole fractions.

This definition is easy to visualize on the molecular basis; it means that there is no difference in interaction *energy* between solute and solvent ($\Delta H_m = 0$) and that the entropy change arises only from the randomness produced by mixing. From Equations (2.25) and (2.26), we have, for the free energy of mixing,

$$\Delta G_m = \Delta H_m - T \Delta S_m = RT \sum_i n_i \ln X_i \qquad (2.27)$$

for an ideal solution. But in general, the free energy of mixing is defined by

$$\Delta G_m = G_{\text{(solution)}} - \sum_i G_{i\text{(pure components)}} \qquad (2.28)$$

The free energy of a particular pure component will equal the number of moles of that component times a free energy per mole. We may regard this free energy per mole of a pure substance as a reference chemical potential, μ_i^0. Then from Equations (2.28) and (2.10)

$$\Delta G_m = \sum_i n_i \mu_i - \sum_i n_i \mu_i^0 \qquad (2.29)$$

or by Equation (2.27)

$$\sum_i n_i(\mu_i - \mu_i^0) = RT \sum_i n_i \ln X_i \qquad (2.30)$$

Since the components of a solution are independently variable, we must have

$$\mu_i - \mu_i^0 = RT \ln X_i \qquad (2.31)$$

That is, the chemical potential will depend on the logarithm of the concentration. In the *standard state* or *reference state*, which has been chosen in this case to be pure component i, $\mu_i = \mu_i^0$. Although the mole fraction scale is satisfactory for describing the behavior of the solvent in a solution of macromolecules, one usually does not employ this concentration scale for the solute. We are interested in solute components that are present only at very low concentrations, and therefore the pure component as a standard state is inconvenient. At low concentrations, a weight or molar concentration will be proportional to mole fraction. Since it is the logarithm of the concentration that occurs in Equation (2.31), any proportionality constant can be absorbed in a redefined μ_i^0. Thus we may equally well write for the solute in dilute, ideal solutions

$$\mu_i = \mu_i^0 + RT \ln C_i \qquad (2.32)$$

where C_i denotes grams per liter. The value of μ_i^0, as well as the definition of the standard state, is understood to depend on the concentration scale used. In this case μ_i^0 is the chemical potential of the ideal solute at unit concentration, for example, when $C_i = 1$ g/liter.

Nonideality can be taken care of in a purely formal way by inserting an activity coefficient. Thus the analog of Equation (2.32) for a nonideal solution can be written as

$$\mu_i = \mu_i^0 + RT \ln C_i y_i \qquad (2.33)$$

where y_i is a function that describes all deviations from ideality. In general, y_i will be a function of T, P, and *all* of the solute concentrations. We expect that, in general, solutions will approach ideal behavior as $C \rightarrow 0$; thus

$$\lim_{C \to 0} y_i = 1 \qquad (2.34)$$

where C is the total concentration of *all* solutes in the solution.

In discussing some properties of macromolecular solutions (for example, osmotic pressure) we shall be interested primarily in the chemical potential of the solvent. For dilute solutions it is most convenient to take the standard state of the solvent as pure solvent and to use the mole fraction scale. Then, we write, in place of Equation (2.31),

$$\mu_1 = \mu_1^0 + RT \ln X_1 f_1 \qquad (2.35)$$

where f_1 is an activity coefficient on the mole fraction scale. But since $X_1 = 1 - X_2$ (where X_2 is the mole fraction solute) and $\ln(1 - X_2) \simeq -X_2 - X_2^2/2 \cdots$,

$$\mu_1 - \mu_1^0 \simeq -RT(X_2 + \cdots) + RT \ln f_1 \qquad (2.36)$$

Now we may wish to express the solute concentration in terms of C_2 rather than

X_2. For dilute solutions

$$X_2 \simeq \frac{C_2 V_1^0}{M_2} \tag{2.37}$$

where V_1^0 is the molar volume of pure solvent. Therefore,

$$\mu_1 - \mu_1^0 = -\frac{RTV_1^0 C_2}{M_2} + RT \ln f_1 \tag{2.38}$$

Both $\ln f_1$ and its first derivative must approach zero as $C_2 \to 0$. Furthermore, we shall (in most cases) expect f_1 to be less than unity (reduced solvent activity) at finite C_2. Therefore, we might expect to be able to represent $\ln f_1$ by a negative power series in C_2, starting with C_2^2:

$$\ln f_1 = -(\alpha C_2^2 + \beta C_2^3 + \cdots) \tag{2.39}$$

where the α, β, and so forth, are unknown constants. This gives, when combined with Equation (2.38),

$$\mu_1 - \mu_1^0 = -RTV_1^0 \left(\frac{C_2}{M_2} + BC_2^2 + \cdots \right) \tag{2.40}$$

where $B = \alpha/V_1^0$. This is known as the *virial expansion* of the solvent chemical potential. The quantities $1/M_2$, B, and so forth, are known as the first, second, and higher *virial coefficients*. The quantity B, the second virial coefficient, serves as a convenient measure of solution nonideality. If $BC_2 \simeq 0$, the higher terms in the virial expansion will presumably be even smaller, and we may write

$$\mu_1 = \mu_1^0 - \frac{RTV_1^0 C_2}{M_2} \tag{2.41}$$

This conforms to our intuitive expectation that the chemical potential of the solvent should be reduced by adding a bit of solute.†

2.2 Solutions of Macromolecules

The solution behavior of biological macromolecules is determined principally by three factors: They are large, they may interact strongly with the solvent or one another, and they frequently carry electrical charges. We shall consider each of these sources of the nonideality, but first it is necessary to say something about the conformations that a macromolecule may exhibit in solution.

The macromolecules encountered by the biochemist (principally proteins, nucleic acids, and polysaccharides) can exhibit a wide variety of solution conformations. This is because these are all polymers, having considerable flexibility in the molecular chain, and in most cases strong intramolecular interactions between side groups. Nevertheless, their conformations can be roughly grouped into three main categories.

†Note that a very small second virial coefficient will exist even for an ideal solution if we retain the first term that was discarded in deriving Equation (2.36). Work this out and compare it to the values in Table 2.1 to see that it is negligible.

Random Coils. A linear polymer that exhibits little resistance to rotation about the bonds in the chain and in which there is little side-group interaction will tend to adopt the kind of random conformation depicted in Figure 2.1(*a*).

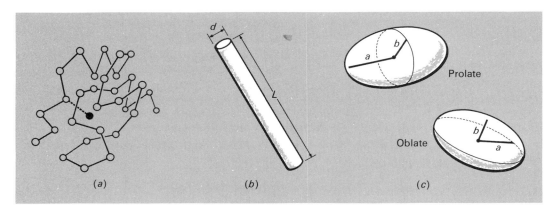

Figure 2.1 Three extreme classes of macromolecular conformation. (*a*) Random coil. The root-mean-square distance of the segments from the center of mass black dot) determines the radius of gyration. (*b*) Rod, with length *L* and diameter *d*. (*c*) Prolate and oblate ellipsoids of revolution. The axial ratio is defined as *a/b*. For a sphere, *a/b* = 1.

Such behavior is often observed with synthetic high polymers, some polysaccharides, and proteins or nucleic acids which have been "denatured" by high temperature or solvents which interact strongly with the side chains. There is, in such cases, no unique three-dimensional structure; rather, the molecule is continually being contorted by the impact of surrounding solvent molecules. Thus we can speak only of *average* dimensions for such molecules. The backbone of such a chain will, at any instant, resemble a three-dimensional random walk, rather like the path of a small molecule undergoing Brownian motion (see Chapter 4). The most useful average dimension for such a molecule is called the *radius of gyration*, R_G. If we regard the chain as being made up of N monomer units connected by $N - 1$ links, R_G is defined by

$$R_G^2 = \frac{\sum_i \overline{R_i^2}}{N} \tag{2.42}$$

where $\overline{R_i^2}$ is the average square of the distance of unit i from the center of mass of the molecule.

The quantity R_G gives a rough idea of the volume of solution "occupied" by the extended molecule. This volume will usually be much greater than that which the molecule would occupy if it were shrunk to a compact sphere. Consequently, when a macromolecule such as a protein is denatured, it will expand to effectively enclose a much larger fraction of the total solution volume.

A statistical analysis of the behavior of such a random coil shows that the average distribution of segments about the center of mass will be spherically symmetric, with an approximately Gaussian distribution of segment density about the center. For a true random coil, R_G will be proportional to the square root of the number of segments [see Tanford (1961)].

Rodlike Macromolecules. Some biopolymers, such as polynucleotides and helical polypeptides, adopt a rodlike conformation in solution. Molecules of this type usually have a simple helical secondary structure. Their solution properties are largely dominated by the length of the rod, the diameter playing a lesser role [Figure 2.1(*b*)].

Globular Macromolecules. Biopolymers in which there are strong side-chain interactions tend to coil up into dense globular conformations. Such are typical of the so-called globular proteins. Such molecules can often be adequately represented, for consideration of their solution properties, as either spheres or ellipsoids of low axial ratio. This kind of model will be used in a number of the following chapters. Figure 2.1(*c*) depicts both prolate (cigar-shaped) and oblate (flattened-sphere) ellipsoids of revolution.

It must not be assumed that such models exactly represent real macromolecules. (For a comparison to a real protein molecule, see Chapter 11.) Nevertheless, for physical calculations, such approximate models frequently work very well.

We now return to a discussion of nonideal behavior in solutions of macromolecules. In the first place, many macromolecules interact strongly with the solvent or with one another. This can have two effects: Either $\Delta H_m \neq 0$, or ΔS_m may contain contributions from ordering or disordering of the solvent. While quantitative analysis is difficult, it is not hard to present a qualitative picture of the effects, at least insofar as the ΔH_m contributions are concerned. A negative heat of mixing corresponds to favorable solute–solvent interation. This should lead to an even greater decrease in the solvent chemical potential than expected for ideal solutions—and thus a positive value of B [see Equation (2.40) and Figure 2.2]. Conversely, if solute–solute interaction is strong, the virial coefficient will be negative. Too large a negative value of B is not to be tolerated, however, for if the value of B is so large that a minimum occurs in the μ_1 versus X_1 graph, phase separation must occur. (See the upper curve in Figure 2.2.) In this case there are two concentrations at which the solvent has the same value of the chemical potential; this means that two liquid phases, of these two compositions, can coexist, for solvent can be transferred from one to the other with zero free-energy change. Solutions corresponding to the broken region of the curve are at best metastable, and should separate into two phases.

The most general cause for nonideality in solutions of macromolecules arises from the sheer size of the particles. It will be recalled that an ideal solution is one for which the ideal entropy of mixing law holds. The derivation of that

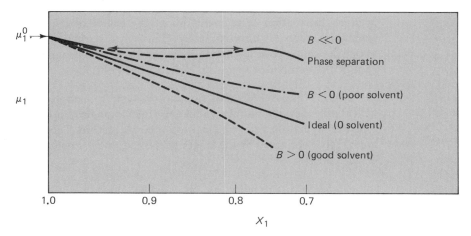

Figure 2.2 A schematic diagram of the behavior of the solvent chemical potential near $X_1 = 1$ for ideal and nonideal solutions. For $B \ll 0$, phase separation will occur, yielding the phases given by the ends of the arrow.

law (see Chapter 1) involved the assumption that the solute molecules are of the same order of size as the solvent molecules, so that solvent and solute might be interchanged at random in a hypothetical lattice. But this is by no means true for macromolecules. Each macromolecule is many times larger than a solvent molecule; in fact, the monomer units of the macromolecule are more comparable to the solvent molecules in size. In other words, a macromolecular solution more nearly resembles one in which the solute particles (the monomer units) are required to move together in clumps.

Put another way, this says that the distribution of solute molecules in a macromolecular solution can never be entirely random. The center of each molecule is excluded from a volume determined by the volumes occupied by all of the other molecules. It is not surprising, then, to find that the nonideality, and hence the virial coefficient, depend on the molecular excluded volume (u) [for a detailed derivation, see Tanford (1961)]. We find that

$$B = \frac{\mathfrak{N}u}{2M_2^2} \tag{2.43}$$

where $\mathfrak{N}$ is Avogadro's number. The excluded volume is determined by the actual molecular dimensions.

For compact spheres: $\quad u = \frac{8M_2 v_2}{\mathfrak{N}} \qquad B = \frac{4v_2}{M_2} \tag{2.44}$

For rods: $\quad u = \frac{2LM_2 v_2}{\mathfrak{N}d} \qquad B = \frac{L}{d}\frac{v_2}{M_2} \tag{2.45}$

where v_2 is the specific volume, L the rod length, and d its diameter. For random coils, we may relate u to the radius of gyration.

$$\text{For random coils:} \quad u = \frac{32\pi\gamma^3 R_G^3}{3} \tag{2.46}$$

where γ is a number close to unity.

Since the R_G of a true random coil is proportional to $M^{1/2}$ (see Chapter 4), we have

$$B = KM_2^{1/2} \tag{2.47}$$

where K is a proportionality constant which depends on the nature of the particular macromolecule. In Table 2.1 are shown representative values of B and $(1 + BM_2C_2)$ for $C_2 = 5$ mg/ml, a concentration in the range frequently employed in the study of macromolecules. The quantity $(1 + BM_2C_2)$ represents a correction term for nonideality. The amount by which it differs from unity is a direct measure of solution nonideality [see Equation (2.40)]. The general conclusion to be drawn is that while this contribution to nonideality is small for spherical molecules, it can become large for very asymmetric rods or random coils. This is understandable, for the number of ways in which such particles can be packed into a solution is quite limited.

In view of the above, it might seem surprising that solutions of macromolecules can exhibit very nearly ideal behavior under some conditions. For example, for random-coil polymers there is often a particular temperature (called the θ temperature) at which $B = 0$. Qualitatively, the reason for this is the following. The excluded volume effect is always such as to lower the chemical potential of the solvent. It is an entropy effect. On the other hand, in poor solvents the solute-solute interaction may be such as to increase the solvent chemical potential. This is an enthalpic term. There may exist a temperature (the θ temperature) at which these enthalpic and entropic terms exactly cancel. Then the solution will behave ideally. Putting this in mechanistic terms, we regard the excluded volume effect as one that pushes molecules apart (they cannot interpenetrate). This may

TABLE 2.1 *Solution nonideality*

Particle	B^a	$(1 + BM_2C_2)^b$
Sphere, $M = 10^5$	3×10^{-5}	1.015
Rod, $L/d = 100$, $M = 10^5$	7.5×10^{-4}	1.375
Random coil, $M = 10^5$, good solvent	5×10^{-4}	1.250
Random coil, θ solvent	0	1.000

^{a}Order of magnitude, in units of cm^3/g^2-mole.
^{b}Assuming that $C_2 = 5$ mg/ml.

Figure 2.3 Excluded volume (*a*) and inter-molecular attraction (*b*) as determinants of solution behavior.

be compensated for by an interaction of the sort that makes solute molecules clump together (see Figure 2.3).

The discovery of a θ temperature for a particular macromolecule–solvent system is very useful for it greatly simplifies all physical studies. The much simpler rules for ideal solutions can then be used.

2.3 Membrane Equilibria

Semipermeable membranes play a number of important roles in biochemistry. On the practical side, we employ them in dialysis and membrane filtration for the purification of macromolecular substances. Again, equilibrium across membranes is used to measure the binding of small molecules and ions to large molecules (see the section on dialysis equilibrium below) or to determine molecular weights. In the cell, membranes take on a similar role, serving to parition regions of the cell, and the cell from its surroundings, by barriers that retain some substances and allow others to pass. The principal functional difference between the membranes that we employ in the laboratory and those found in the cell (aside from composition) is that the former are *passive*; that is, they act only as barriers and play no active role in the transport of materials. The *active transport* in some cell membranes is a very different thing. Here, at a free-energy price, materials are selectively transported against concentration differences. More will be said of this later.

In this chapter we shall mainly be concerned with the equilibrium phenomena arising from the existence of passive, semipermeable membranes in a system. We shall first write some general rules, then discuss osmotic pressure and the use of dialysis equilibrium, and then turn to the complications introduced when some of the solutes in the solutions carry an electrical charge. At the close, a little will be said about active transport.

EQUILIBRIUM ACROSS A MEMBRANE

Consider a solution that is separated from solvent by a semipermeable membrane, as in Figure 2.4. We shall assume that solvent molecules and some low-molecular-weight solute components can pass through the membrane,

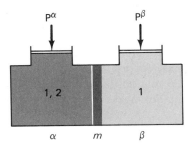

Figure 2.4 Two phases, α and β, are separated by a semipermeable membrane m. The membrane is permeable to component 1 but not to component 2. We can adjust the pressure on α and β as we like.

whereas some kinds of large solute molecules cannot. In practice, the cellophane membranes used most commonly in biochemical laboratories pass materials of $M < 10{,}000$ and retain larger molecules, but membranes can be prepared that select at much higher or much lower molecular weight.

The system must be regarded as consisting of two phases (α and β), for the fact that the large molecules are restricted to one side means that the system cannot be homogeneous throughout. What are the requirements for equilibrium? We can write them immediately from Equation (2.18):

$$\mu_i^\alpha = \mu_i^\beta \tag{2.48}$$

which holds for all substances that can pass through the membrane. For those substances that cannot pass through, we can make no corresponding statement. Equation (2.48) is the starting point for all discussions of membrane equilibria. For a system composed, for example, of hemoglobin, oxygen, buffer salts, and water, Equation (2.48) would apply to each of the components except hemoglobin.

OSMOTIC PRESSURE

Let us consider the implications of Equation (2.48) in a simple system involving only a solvent (component 1) that can pass the membrane and a solute (component 2) that cannot. On one side of the membrane (α) we have both 1 and 2; on the other side (β) we have only the solvent, component 1. Both sides are initially assumed to be at the same T and P. The only equation we have is the statement that

$$\mu_1^\alpha = \mu_1^\beta \tag{2.49}$$

at equilibrium. However, this seems to lead to a strange conclusion. According to Equation (2.41), the chemical potential of solvent in a solution will be less than that in pure solvent ($\mu_1 - \mu_1^0 < 0$). It would appear then, that equilibrium will not be attainble, for no matter how much solvent is transferred into phase α, the inequality will still exist. This is actually the case in the system as we have set it up; it is not generally possible for a solution to be in equilibrium with pure solvent at the same T and P. The entropy can always be increased (and the free energy decreased) by transferring some more solvent to further dilute the solution.

To attain equilibrium, we must evidently produce some *other* difference on the two sides of the membrane, so as to produce a chemical potential difference that will compensate for the difference produced by the presence of solute on one side. We might imagine, for example, having the two sides at different temperature. However, it seems impossible to conceive of a membrane that would transmit matter but not heat.† This leaves us only pressure to work with.

Now it is clear that the chemical potential, like any free-energy quantity, will depend on the pressure exerted on a system. Since a membrane can support a pressure difference, suppose we try this. An apparatus we might use is shown in Figure 2.4. We shall keep the pressure on the solvent side (β) at some nominal value, P_0 (1 atm, for example) and increase the pressure on side α (the solution side) to $P_0 + \pi$. What effect will this have on μ_1^α? We know that the variation of free energy with pressure is given by

$$\left(\frac{\partial G}{\partial P}\right)_T = V \tag{2.50}$$

which can be obtained directly from Equation (1.48). We can write for the partial molar free energy (μ_1) the analogous result

$$\left(\frac{\partial \mu_1}{\partial P}\right)_T = \bar{V}_1 \tag{2.51}$$

where $\bar{V}_1$ is the partial molar volume of solvent. Then we have

$$\mu_1^\beta = \mu_1^0 \tag{2.52}$$

(since on side β we have pure solvent at standard conditions) and

$$\mu_1^\alpha = \mu_1^0 - RTV_1^0\left(\frac{C_2}{M_2} + BC_2^2 + \cdots\right) + \int_{P_0}^{P_0+\pi} \bar{V}_1 \, dP \tag{2.53}$$

Assuming that $\bar{V}_1 = V_1^0 =$ constant $=$ the molar volume of pure solvent at 1 atm (a good approximation for low pressures and dilute solutions), we obtain, from Equations (2.52) and (2.53),

$$\mu_1^0 = \mu_1^0 - RTV_1^0\left(\frac{C_2}{M_2} + BC_2^2 + \cdots\right) + V_1^0\pi$$

or

$$\pi = RT\left(\frac{C_2}{M_2} + BC_2^2 + \cdots\right) \tag{2.54}$$

We have calculated the pressure difference required to equate the chemical potential of solvent on the two sides of the membrane. This we call the *osmotic pressure*. It must be applied for the system to be at equilibrium. If the solution

---

†If we are willing to accept conditions that only approach equilibrium, this can be done. Thus one type of "osmometer" makes use of the temperature difference established between two droplets, one containing only solvent and the other solution.

is very dilute or if $B = 0$, we obtain

$$\pi = \frac{RTC_2}{M_2} \tag{2.55}$$

This equation shows why π is important; it allows us to measure the molecular weight. Since we shall have to use solutions of finite concentration to measure π and since these may be nonideal, a practical equation for the calculation of M_2 can be written as

$$\frac{\pi}{C_2} = \frac{RT}{M_2} + BRTC_2 + \cdots \tag{2.56}$$

Graphs of osmotic pressure data for globular and unfolded proteins are shown in Figure 2.5. It is evident that extrapolation to $C_2 = 0$ is necessary for accurate results. Some representative data are given in Table 2.2.

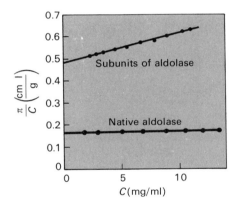

Figure 2.5 Osmotic pressure data for the protein aldolase in buffer at neutral pH and the subunits of this protein in 6 M guanidine hydrochloride. Molecular weights are given in Table 2.2. Note the larger virial coefficient for the unfolded subunits. Data from F. J. Castellino and O. R. Barker, *Biochemistry*, **7**, 2207 (1968). Reprinted by permission of the American Chemical Society.

TABLE 2.2 *Some molecular weights obtained from osmotic pressure studies*

Substance	M_n
Ovalbumin	44,600
Hemoglobin	66,500
Aldolase	156,500
Aldolase subunits	42,400
Amylose, various samples	32,000–150,000

One can measure the osmotic pressure in a number of ways. For example, each side of the apparatus depicted in Figure 2.4 could be fitted with a standpipe, and solvent could be allowed to flow from β to α until the hydrostatic pressure difference equaled π. Alternatively, we could adjust the pressure difference until no flow of solvent occurred. The latter method is used in modern, automated osmometers.

Equation (2.56) shows how the molecular weight of a homogeneous solute can be measured. Now suppose the solute is heterogeneous, that is, that it consists of a mixture of macromolecular components. What result is obtained then? At very low concentrations the osmotic pressure can be regarded as the sum of individual contributions:

$$\pi = \sum_i \pi_i \tag{2.57}$$

If

$$\pi_i = \frac{RTC_i}{M_i} \tag{2.58}$$

we obtain

$$\pi = RT \sum_i C_i/M_i \tag{2.59}$$

This does not look very useful. But suppose we divide and multiply the right by $\sum_i C_i = C$, the total solute concentration. Then

$$\pi = \frac{RTC}{\sum_i C_i / \sum_i (C_i/M_i)} \tag{2.60}$$

This looks like Equation (2.55). If we define an *average molecular weight, M_n*, by

$$M_n = \frac{\sum_i C_i}{\sum_i (C_i/M_i)} \tag{2.61}$$

we have

$$\pi = \frac{RTC}{M_n} \tag{2.62}$$

or, for finite concentrations,

$$\pi = RT\left(\frac{C}{M_n} + \bar{B}C^2 + \cdots\right) \tag{2.63}$$

where $\bar{B}$ is an average virial coefficient. These important results tell us that the osmotic pressure measures an average of the molecular weights in a heterogeneous solute and how that average is defined. We shall encounter average molecular weights in many situations. A number of useful averages are tabulated in Table 2.3, expressed in terms of both weight and number concentrations. Note that the average measured by osmotic pressure is just the mean molecular weight, on a number concentration basis. This average will be obtained from any colligative property (such as osmotic pressure, freezing-point depression, or boiling-point elevation) that depends on the number of solute molecules per unit volume. Similar statements can be made about the other molecular weight averages listed in Table 2.3. Note, for example, that the weight-average molecular weight is the mean when components are counted by the *weight* of each. This average will be given by any method, such as sedimentation equilibrium (Chapter 5) or light scattering (Chapter 9), which measures the weight of each component. It should also be noted that these different averages will not be numerically the same for a mixture of components. Those which weight larger molecules more heavily (such as the weight and Z averages) will be numerically larger than the number average. Only for a homogeneous macromolecule will methods that yield different averages give the same value for the molecular weight.

TABLE 2.3 *Average molecular weights*

	Definition[a]	
Average	*In Terms of N_i*	*In Terms of C_i*
Number average, M_n	$\sum_i N_i M_i \big/ \sum_i N_i$	$\sum_i C_i \big/ \sum_i (C_i/M_i)$
Weight average, M_w	$\sum_i N_i M_i^2 \big/ \sum_i N_i M_i$	$\sum_i C_i M_i \big/ \sum_i C_i$
Z average, M_z	$\sum_i N_i M_i^3 \big/ \sum_i N_i M_i^2$	$\sum_i C_i M_i^2 \big/ \sum_i C_i M_i$

[a]C_i is the concentration in weight per volume; N_i is the concentration in numbers of moles or molecules per unit volume.

DIALYSIS EQUILIBRIUM

Frequently, use is made of the membrane equilibrium to measure the binding of small molecules or ions to a macromolecule. The principle is easily visualized. The macromolecular solution is placed inside a membrane bag (phase α) and suspended in a solution containing the small molecules (phase β) (Figure 2.6). At equilibrium, any excess in concentration of the low-molecular-

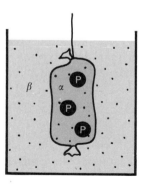

Figure 2.6 Dialysis equilibrium. The small dots represent solute molecules that can pass through the membrane. If some are bound by the protein P, the total inside concentration of these molecules will be greater than that outside.

weight substance on the macromolecular side of the membrane is taken as evidence for binding. Although this appears straightforward, a closer analysis is easy. Calling the low-molecular-weight substance (which we presume to be a nonelectrolyte) component 2, we have at equilibrium

$$\mu_2^\alpha = \mu_2^\beta \tag{2.64}$$

Neglecting the very small difference in chemical potential resulting from the equilibrium pressure difference between the two sides, we obtain, from Equation (2.33),

$$RT \ln C_2^\alpha y_2^\alpha = RT \ln C_2^\beta y_2^\beta \tag{2.65}$$

or

$$C_2^\alpha y_2^\alpha = C_2^\beta y_2^\beta \tag{2.66}$$

Note that to this point we have made no distinction between bound and unbound component 2 on side α. However, if we assume that there exists, on side α, both bound and unbound component 2, and that the environments are sufficiently similar that the activity coefficient for free component 2 is the same on both sides, we have

$$C_2^\alpha(\text{unbound}) = C_2^\beta \tag{2.67}$$

and thus

$$C_2^\alpha(\text{bound}) = C_2^\alpha - C_2^\beta \tag{2.68}$$

It is clear that a number of somewhat arbitrary assumptions have gone into this result. It is not surprising, therefore, that the numbers obtained from such analysis can even be such as to indicate negative binding. In fact, a simple mechanical picture leads to the expectation of such results. Suppose that the macromolecule does not bind the solute at all but simply excludes it from the volume occupied by the macromolecules. Then $C_2^\alpha < C_2^\beta$, and we would interpret this as negative binding. It is clear that unless binding is very strong, analyses depending on the assumptions above must be treated with caution.

EFFECTS OF SOLUTE CHARGE: THE DONNAN EQUILIBRIUM

So far, we have assumed that the substances involved in membrane equilibria are uncharged. If this is not the case and the macromolecules are polyelectrolytes, the situation becomes more complicated. Since polyelectrolyte behavior is the rule rather than the exception with biopolymers, such complications are by no means trivial. All proteins and nucleic acids, for example, are polyelectrolytes. In this section we shall consider only the behavior of strong polyelectrolytes and for concreteness will take the case of a polymer that dissociates according to the rule

$$PX_z \longrightarrow P^{+z} + zX^- \tag{2.69}$$

where X^- is a monovalent anion. The general result is symmetric to charge type, so we do not lose generality by taking the macromolecular species P^{+z} to be a polycation.

If the polycation and its counterions are the only electrolytes present in the system, the result is trivial. Since the counterions must remain on the same side of the membrane as the polymer (to maintain electrical neutrality), we simply have $z + 1$ particles present for every solute molecule introduced. All colligative properties, including osmotic pressure, should then be increased by a factor of $z + 1$, and the material will behave as if it has a molecular weight $1/(z + 1)$ of the undissociated polymer. If our aim was to measure the molecular weight of the polycation, we would be grossly in error. We have taken the number-average molecular weight of the mixture of polycations and counterions.

Fortunately, this rather disastrous situation is rarely approached, for polyelectrolytes are almost always studied in the presence of some additional low-molecular-weight electrolyte, and this markedly changes the results. Let us consider, then, the more complex system shown in Figure 2.7, in which a 1:1 electrolyte ($BX \longrightarrow B^+ + X^-$) has been added to the system. The membrane is

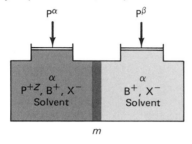

Figure 2.7 Membrane equilibrium with a polyelectrolyte solute.

assumed to be permeable to B^+, X^-, and water but impermeable to P^{+z}. If this is true, we must have

$$\mu_{BX}^\alpha = \mu_{BX}^\beta \tag{2.70}$$

We shall assume the solution to be ideal. The chemical potential of the salt can then be written as $\mu_{BX} = \mu_{BX}^0 + RT \ln c_B c_X$, where c_B and c_X are the molar concentrations of B and X. Then, neglecting the difference $\mu_{BX}^{0\alpha} - \mu_{BX}^{0\beta}$, we have, from Equation (2.70),

$$c_B^\alpha c_X^\alpha = c_B^\beta c_X^\beta \tag{2.71}$$

A second equation is obtained from the requirement for electrical neutrality:

On Side α On Side β

$$zc_P^\alpha + c_B^\alpha = c_X^\alpha \qquad c_B^\beta = c_X^\beta \tag{2.72}$$
$$+ \qquad - \qquad + \qquad -$$

where c_P is the molar concentration of polymer.

We now combine Equations (2.71) and (2.72) to give

$$(c_B^\beta)^2 = (c_B^\alpha)^2 \left(1 + \frac{zc_P}{c_B^\alpha}\right) \tag{2.73}$$

$$(c_X^\beta)^2 = (c_X^\alpha)^2 \left(1 - \frac{zc_P}{c_X^\alpha}\right) \tag{2.74}$$

or,

$$c_B^\beta = c_B^\alpha \left(1 + \frac{zc_P}{c_B^\alpha}\right)^{1/2} \tag{2.75}$$

$$c_X^\beta = c_X^\alpha \left(1 - \frac{zc_P}{c_X^\alpha}\right)^{1/2} \tag{2.76}$$

This states the result we might have anticipated; B^+ is more concentrated in phase β than in phase α, whereas X^- is more concentrated in phase α, since it is the counterion to PX^{z+}. These concentration differences in the ionic species will contribute to the osmotic pressure. We have assumed ideal behavior, so that the total osmotic pressure may be written (in terms of molar concentrations) as

$$\pi = RT[c_P^\alpha + (c_B^\alpha - c_B^\beta) + (c_X^\alpha - c_X^\beta)] \tag{2.77}$$

Simple expressions for $c_B^\alpha - c_B^\beta$ and $c_X^\alpha - c_X^\beta$ can be obtained from Equations (2.75) and (2.76) only if the quantities zc_P/c_B^α and zc_P/c_X^α are small compared to unity. Then we can expand these equations by the binomial theorem and retain only the first few terms:

$$c_B^\beta = c_B^\alpha \left[1 + \frac{zc_P}{2c_B^\alpha} - \frac{z^2 c_P^2}{8(c_B^\alpha)^2} + \cdots \right] \tag{2.78}$$

$$c_X^\beta = c_X^\alpha \left[1 - \frac{zc_P}{2c_X^\alpha} - \frac{z^2 c_P^2}{8(c_X^\alpha)^2} - \cdots \right] \tag{2.79}$$

Rearranging, and combining the terms, we find that those in c_P cancel, whereas the quadratic terms do not.

$$(c_B^\alpha - c_B^\beta) + (c_X^\alpha - c_X^\beta) \simeq \frac{z^2 c_P^2}{8} \left(\frac{1}{c_B^\alpha} + \frac{1}{c_X^\alpha} \right) + \cdots \tag{2.80}$$

The cubic term will nearly cancel, so the result given is in many cases a very good approximation. We can further approximate by noting that

$$\frac{1}{c_B^\alpha} + \frac{1}{c_X^\alpha} \simeq \frac{2}{c_{BX}} \tag{2.81}$$

where c_{BX} is the concentration of low-molecular-weight electrolyte added. The approximation is quite good, since although phase α is a bit richer in X^-, it is a bit more dilute in B^+, so the deviations nearly cancel. The final result is

$$\pi = RT \left(c_P + \frac{z^2 c_P^2}{4c_{BX}} + \cdots \right) \tag{2.82}$$

Now writing the macroion concentration in grams per liter rather than moles per liter, we obtain

$$\pi = RT \left(\frac{C_P}{M_P} + \frac{z^2 C_P^2}{4M_P^2 c_{BX}} + \cdots \right) \tag{2.83}$$

Thus we end with the rather remarkable result that the Donnan effect produces a "pseudo-nonideality." The correction is given approximately by a quadratic term in the macromolecule concentration, just as are the other non-ideality corrections we have considered. Note that this term varies as the square of the charge/mass ratio; it will rapidly become a serious correction as this ratio increases. On the other hand, the effect of the added electrolyte is very evident; the magnitude of the correction term is inversely proportional to that concentration.

Note, however, that Equation (2.83) indicates that as long as a sufficient concentration of electrolyte is present to make the approximations valid, the correct molecular weight for the macroion will be obtained by extrapolation to infinite dilution.

The Donnan effect is by no means confined to membrane equilibria. Similar effects are observed in any experiment in which there is a tendency to produce a separation of ionic species; sedimentation is one example (see Chapter 5). The effect has its origin in the tendency of small counterions to diffuse away from local concentrations of macroions, a tendency countered by the requirement that electrical neutrality be maintained throughout the solution. Such apparent nonideal behavior will generally be observed with polyelectrolytes, but it is not the sole cause of nonideality in such solutions. The mutual attraction or repulsion of polyelectrolytes of unlike or like sign as well as the excluded volume effect will provide additional contributions to the virial coefficient. However, especially in solutions dilute in supporting electrolyte, the Donnan effect is often dominant. It can always be decreased, however, as Equation (2.83) shows, by increasing the concentration of "supporting electrolyte."

2.4 Transport Across Biological Membranes

Passage of small molecules and ions across cellular membranes is of the utmost importance, both for the interaction of a cell with its environment and for the redistribution of substances among cellular compartments. Equally important are the control and selectivity of this process, for certain materials must not be allowed to pass and others (such as some nutrients) are encouraged to cross the membrane even *against* a concentration gradient.

Three modes of material exchange across biological membranes are recognized: *passive*, *facilitated*, and *active transport*.

Passive Transport. This is nothing more than diffusion across the membrane. It can be a functionally useful process only for those substances which are appreciably soluble in the membrane itself. As is shown in Chapter 4, the steady state of diffusion will correspond to an approximately linear concentration gradient in the membrane, producing a flow across the membrane proportional to the chemical potential difference between the two sides. At equilibrium the chemical potential will become the same on both sides, and diffusion will cease. It is important to note that it is the chemical potential, not necessarily the concentration, which is equalized at equilibrium. In many cases macromolecules within a cell may bind appreciable quantities of an ion or other substance. Analysis of the cellular contents for this component would reveal a larger total concentration inside than outside at equilibrium, even though the chemical potential were the same within and without. The situation is exactly analogous to the equilibrium dialysis experiment described in the preceding section. In evaluating the significance of the concentration difference for ions at equilibrium, the Donnan effect must also be taken into account.

Facilitated Transport.   Often the solubility of an ion or hydrophilic molecule in a lipid membrane is so low that passive transport would be exceedingly slow. Such ions or molecules are often carried across cellular membranes by the use of *carrier* molecules. Such a carrier will have one or more binding sites for the ligand in question, and will at the same time be freely soluble in the membrane lipid. An example is shown in Figure 2.8. The molecules' cagelike interior nicely accommodates a potassium ion; its hydrophobic exterior confers membrane solubility. The net effect of facilitation of transport is simply to increase its rate; the final state, as in passive diffusion, is an equalization of chemical potential.

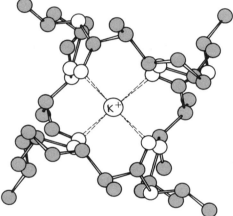

Figure 2.8 Valinomycin is an antibiotic that facilitates transport of potassium ions through membranes. The antibiotic is a cyclic structure; its external surface is hydrophobic, but its interior forms a chelating "cage" to hold a potassium ion. After Lehninger, *Principles of Brochemistry*, Worth Publishers, Inc., New York, 1982. Reproduced by permission of the author and the publisher.

Active Transport.   From what has been said in the preceding sections, it would seem incredible that a membrane could spontaneously and selectively transport molecules or ions so as to *produce* large concentration differences. Yet this is so, and even the example used in Chapter 1, the diffusion of sugar out of a dialysis bag, can be neatly reversed by certain bacteria that can spontaneously transport sugar inward through their cell walls even though the outside concentration is much less than that within the bacteria. How can this be so? Does the second law not always hold? For surely we have shown that the free-energy change accompanying the transport of material against a concentration gradient is positive.

This is, of course, true, but when such processes are observed, we must consider other, more subtle possibilities rather than abandon the second law. A first possibility is suggested by the earlier discussion of dialysis equilibrium. It is actually a gradient in chemical potential that determines the spontaneous direction of free diffusion, and we may be observing a situation in which most of the solute on the high-concentration side of the membrane is actually complexed or bound in some way. If this fact is not known (and it is not always easy to detect such binding in a complicated mixture), we may think that the concentration of free solute is much higher than it really is; the free solute may actually be more dilute than on the "low-concentration" side of the membrane. Such situations are by no means rare; the scavenging of trace metals from sea water by organisms often operates by such mechanisms.

Such behavior is not true active transport. There remains a class of phenomena in which an actual chemical potential difference is established and maintained. But even these involve no violation of the laws of thermodynamics. The clue is found in the fact that actively metabolizing membranes are required. Evidently, the process of transport across a membrane (which costs a free-energy price when the chemical potential gradient opposes it) can be coupled to a spontaneous reaction that has a larger negative free-energy change.

To take a specific example, consider the *sodium–potassium pump* shown in Figure 2.9. Most cells can maintain a high concentration of K^+ and a low

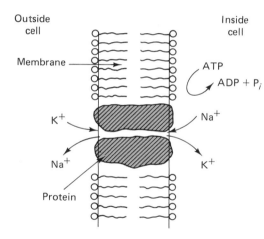

Figure 2.9 A speculative view of the Na^+–K^+ "pump." Phosphorylation of the protein by ATP is coupled to the transport of about three Na^+ ions from the inside to the outside; the completion of the cycle by dephosphorylation and release of phosphate is coupled to the transport of about two K^+ ions into the cell. Presumably, conformational changes in the protein accompanying phosphorylation and dephosphorylation drive these processes.

concentration of Na^+ internally, despite the fact that they are bathed in a medium in which Na^+ is high and K^+ is low. Thus they must be capable of transporting both of these ions against concentration gradients. To examine this phenomenon quantitatively, we must consider a further complication: There is usually an electrical potential gradient across biological membranes like the cell membrane. If an ion, such as the sodium ion in this example, is present in such a field, its chemical potential, as we have defined it so far, does not wholly describe the contribution of these ions to the free energy of the solution. We must include a term for the electrical potential energy. So we write the *electrochemical potential*

$$\mu'_i = \mu^0_i + RT \ln c_i + z_i \mathfrak{F} \mathcal{E} \tag{2.84}$$

where ε is the electrical potential (volts), and $\mathcal{F}$ is the Faraday constant (23.06 kcal/mole-V). We shall encounter an exactly analogous situation in Chapter 5, when we discuss the behavior of macromolecules in *centrifugal* fields.

According to Equation (2.84), the free-energy requirement for transporting a mole of Na^+ ions across the membrane in Figure 2.9 is given by

$$\Delta G = \mu'_{Na^+}(out) - \mu'_{Na^+}(in)$$

$$= RT \ln \frac{c'_{Na^+}(out)}{c'_{Na^+}(in)} + \mathcal{F} \Delta\varepsilon \qquad (2.85)$$

where $\Delta\varepsilon$ is the potential difference across the membrane ($\varepsilon_{out} - \varepsilon_{in}$). To obtain an estimate for ΔG, let us consider a typical situation, where $c(out)/c(in)$ 10, $\Delta\varepsilon = 70$ mV, $T = 37°C$. Calculation yields the value $\Delta G = 3.03$ kcal/mole. This large, positive free-energy change attests to the nonspontaneity of the process. But sodium transport in many cells is carried out by the sodium–potassium pump, in which the transport of Na^+ out of the cell and K^+ into the cell is coupled to the hydrolysis of ATP. Since the hydrolysis of 1 mole of ATP under cellular conditions involves a free-energy change in excess of -7 kcal/mole, it is clear that the overall process (transport plus ATP hydrolysis) can still be spontaneous. In fact, several Na^+ and K^+ are transported for each ATP hydrolyzed. The free-energy cost for transporting K^+ into the cell is small, since the electrical potential gradient is favorable in this case. The tranpsort of these two ions is coupled, so that as Na^+ is pumped out, K^+ is simultaneously pumped in.

Such examples, of which there are many in biological systems, do not involve any violation of the second law. Transport against the gradient continues only as long as it can be coupled to the hydrolysis of ATP. Viewed in this way, active transport is not paradoxical or even surprising. All of the anabolic processes in life, the organizing, segregating, and arranging that goes on, are, taken one by one, nonspontaneous processes. The assembly of a protein molecule from a collection of amino acids involves an enormous free-energy *increase*. But all of these processes are linked and depend ultimately on the utilization of a little of the energy of sunlight to drive them, for a while, in nonspontaneous directions. But the entropy price is always paid on the grand scale. For every algal cell that lives, a little more of the sun's energy is absorbed, the earth heats a bit more, and the energy of the universe is a little more evenly distributed.

PROBLEMS

*1. Using the Gibbs–Duhem equation, obtain an expression for $\mu_2 - \mu_2^0$ from the equation

$$\mu_1 - \mu_1^0 = -RTV_1^0 C_2 \left(\frac{1}{M_2} + BC_2 \right)$$

2. Bushy stunt virus particles are approximately spherical and of radius 140 Å, and $M = 1.07 \times 10^7$. Calculate, on the basis of excluded volume theory, the second virial coefficient of bushy stunt virus.

3. (a) Show that on the basis of excluded volume theory the *percent* deviation from ideal behavior is given by

$$\frac{\mathfrak{N}uC_2}{2M_2} \times 100$$

where C_2 is in g/ml.

On this basis, estimate the percent deviation for a 3-mg/ml solution of a protein of molecular weight 30,000, assuming the following conformations:

(b) A hard sphere, density 1.4 g/cm^3.

(c) An α helix of the same density. You may assume an average residue weight of 110 g/mole, a 1.5-Å rise per residue, and a helix radius of 3 Å.

4. Suppose that a protein of molecular weight M dissociates so as to yield one each of N subunits of weights $M_1, M_2, M_3, \ldots, M_N$. Show that the number-average molecular weight of the resulting mixture is just M/N.

5. The following osmotic pressure data have been obtained for lactate dehydrogenase in two different buffer systems:

(a) In 0.1 M KCl/0.1 M potassium phosphate buffer at 25°C (buffer density = 1.012 g/ml), π/c had the value of 0.183 cm of solvent/(g/liter), independent of concentration. Calculate M_n.

(b) In 6.0 M guanidine hydrochloride, a denaturing and dissociating solvent, with density 1.150 g/ml, the following data were obtained:

c (g/liter)	π/c [cm Solvent/(g/liter)]
1.25	0.59
2.50	0.60
3.75	0.61
5.00	0.62

Calculate M_n under these circumstances, and estimate the number of subunits in a native lactate dehydrogenase molecule.

6. The following data describe the osmotic pressure (in centimeters of solution) of solutions of a fraction of polyvinyl chloride in methyl ethyl ketone (MEK) at 25°C. The density of MEK is 0.80 g/ml, and the solution may be assumed to be of equal density. Calculate $\overline{M}_n$ and the second virial coefficient B for the sample.

C (g/100 cm^3)	π (cm Solution)
0.167	0.40
0.489	1.44
0.654	2.10
0.801	2.76
0.980	3.83

7. Given a double-stranded DNA molecule, of molecular weight 66,000, containing 100 base pairs of DNA. Such a molecule will behave like a rigid rod, of diameter 25 Å and length 340 Å.

(*a*) Estimate B and $1 + BM_2C$ for a 1-mg/ml solution, assuming that every phosphate residue is charged. Assume that the excluded volume term and the Donnan term are additive to give the total B.

(*b*) Repeat the calculation, with the more realistic assumption that about 80% of the charge will be neutralized by condensed counterions.

8. Calculate the free-energy change involved in transporting 1 mole of sucrose from a region where the concentration is 0.001 M to a region where it is 1 M. Theoretically, could this be accomplished by the hydrolysis of 1 mole of ATP in a linked process? State your assumptions.

*9. Show that a molecule which has a weak tendency to dimerize behaves as if it had a negative virial coefficient.

REFERENCES

General

Tanford, C.: *Physical Chemistry of Macromolecules*, John Wiley & Sons, Inc., New York, 1961, Chap. 4. An excellent and comprehensive treatment of solution theory, especially as applied to macromolecules.

Special Topics

Schultz, S. G.: *Basic Principles of Membrane Transport*, Cambridge University Press, Cambridge, 1980. A brief but up-to-date monograph on all aspects of membrane transport.

Tombs, M. P. and A. R. Peacocke: *The Osmotic Pressure of Biological Macromolecules*, Clarendon Press, Oxford, 1974.

Chapter Three

CHEMICAL EQUILIBRIUM

Biochemistry is concerned with many types of chemical equilibria. Some of these involve the small molecules metabolized and utilized as monomers in construction of the cell. Other important equilibria involve macromolecules, either in reactions with other macromolecules or in reactions with small molecules. All of these reactions occur in solution, and the usual approach of physical chemistry texts, to emphasize gas-phase reactions, is of little direct interest to the biochemist. In this chapter we shall discuss solution equilibria in general and then treat in detail the general class of multiple equilibria involving macromolecules and small molecules. Brief discussions of order–disorder equilibria and macromolecular association reactions will complete the chapter.

3.1 Thermodynamics of Chemical Reactions in Solution

As emphasized in Chapter 2, discussion of the thermodynamics of solutions centers on the chemical potential. Suppose that we have a general chemical reaction

$$a\text{A} + b\text{B} + \cdots \; \rightleftharpoons \; g\text{G} + h\text{H} + \cdots \tag{3.1}$$

in which a moles of A, b moles of B, and so forth, at molar concentrations c_A, c_B, and so forth, are converted into g moles of G, h moles of H, and so forth, at concentrations c_G, c_H, and so forth.† We may write the free-energy change, ΔG, as

$$\Delta G = G(\text{final state}) - G(\text{initial state}) \tag{3.2}$$

If the reaction occurs at constant T and P, Equation (2.10) yields

$$\Delta G = g\mu_G + h\mu_H + \cdots - a\mu_A - b\mu_B - \cdots \tag{3.3}$$

where μ_G, μ_H, and so forth, are the chemical potentials at the concentrations existing in the solution. Recalling that

$$\mu_i = \mu_i^0 + RT \ln c_i y_i \tag{3.4}$$

we may rewrite Equation (3.3) as

$$
\begin{aligned}
\Delta G &= g\mu_G^0 + h\mu_H^0 + \cdots - a\mu_A^0 - b\mu_B^0 - \cdots + RT \ln \frac{(c_G y_G)^g (c_H y_H)^h \cdots}{(c_A y_A)^a (c_B y_B)^b \cdots} \\
&= (g\mu_G^0 + h\mu_H^0 + \cdots - a\mu_A^0 - b\mu_B^0 - \cdots) + RT \ln \frac{c_G^g c_H^h \cdots}{c_A^a c_B^b \cdots} \\
&\quad + RT \ln \frac{y_G^g y_H^h \cdots}{y_A^a y_B^b \cdots}
\end{aligned}
\tag{3.5}
$$

The three terms in this equation may be distinguished as follows: The first, involving the μ_i^0's, is the standard-state free-energy change (ΔG^0). It corresponds to the ΔG that would be observed if a moles of A and so forth in the standard state formed g moles of G and so forth, also in the standard state. The second term, involving the c_i's, gives the effect of the actual initial and final concentrations on the free-energy change. It is corrected by the third term, which involves only the activity coefficients. For purposes of discussion, we may assume this activity coefficient term to be zero; this would be true, for example, if all components were present in such great dilution that they could be considered to behave ideally. In this event, all y_i's will equal unity, and the third term will vanish. We may then rewrite Equation (3.5) as

$$\Delta G = \Delta G^0 + RT \ln \frac{c_G^g c_H^h \cdots}{c_A^a c_B^b \cdots} \tag{3.6}$$

The values of ΔG^0 for various reactions have been tabulated (see Table 3.1 for some examples). Although such data are useful, it should be emphasized that they do *not* represent the ΔG values accompanying reactions under all possible conditions. In fact, since the standard states are usually defined (at least for low-molecular-weight materials) as unit molarity, the conditions for the standard-state reactions are usually ludicrous by biological standards. To make this clear, let us take a common example. The hydrolysis of adenosine triphosphate (ATP) to yield adenosine diphosphate (ADP) and inorganic phosphate (P_i)

---

†Note that we are not assuming equilibrium here; we simply take reactants under stated conditions and convert them to products.

TABLE 3.1 *Standard free-energy changes accompanying some reactions of bochemical importance[a]*

I. Hydrolysis reactions. Based on a standard state of 1 M concentration of reactants (except H⁺, when pH is given). Water activity assumed unity.

Reactant	pH	Other Conditions	$-\Delta G^0$ (kcal/mole)
Phosphoenolpyruvate	7.0		14.8
Carbamyl phosphate	9.5		12.3
3-Phosphoglycerol phosphate	6.9		11.8
Creatine phosphate	7.0	37°C	10.3
ATP ($\longrightarrow$ AMP + PP$_i$)	7.0	Excess Mg²⁺	7.7
ATP ($\longrightarrow$ ADP + P$_i$)	7.0	37°C, excess Mg²⁺	7.3
ATP ($\longrightarrow$ ADP + P$_i$)	7.4	0 Mg²⁺	9.6
Glucose-1-phosphate	7.0	25°C	5.0
Pyrophosphate	7.0	0.005 M Mg²⁺	4.5
Glucose-6-phosphate	7.0	25°C	3.3
α-Glycerophosphate	8.5	38°C	2.2
Sucrose			7.0
Maltose			4.0
Glycogen			4.0
Glycylglycine		37.5°C	3.6
Benzoyltyrosyl glycinamide	6.5	23°C	0.42

II. Heats and free energies of formation for formation of compound in standard state from C (graphite), H$_2$ (gas), O$_2$ (gas), N$_2$ (gas), all in standard states.

Substance	$-\Delta H_f^0$	$-\Delta G_f^0$
Glycine	128.4	90.3
L-Leucine	152.4	83.1
L-Tryptophan	99.2	28.5
Glycylglycine	178.1	116.6
DL-Leucylglycine	205.6	112.4
Glycyl-DL-tryptophan	148.8	54.7
Water (*l*)	68.32	56.69

[a]Most data are from H. Sober (ed.), *The Handbook of Biochemistry and Molecular Biology*, Chemical Rubber Co., Cleveland, Ohio, 1968.

may be written

$$ATP + H_2O \rightleftharpoons ADP + P_i + H^+ \qquad (3.7)$$

so we might write

$$\Delta G = \Delta G^0 + RT \ln \frac{c_{ADP}c_{P_i}}{c_{ATP}} \qquad (3.8)$$

We do not include c_{H^+} because ΔG^0 is defined in this case for pH = 7; c_{H^+} is fixed at 10^{-7} M. Note that the water concentration has also not been included. The solutions are usually so dilute that any water consumed in the reaction represents a negligible change in the total water concentration.† The term $(-RT\ln c_{H_2O})$, like the term $(RT\ln c_{H^+})$, has, in fact, been included in the ΔG^0 term, since it is so very nearly a constant. Now ΔG^0 for this reaction at 25°C is, by the best estimate, about -7 kcal at pH 7. However, in the cell, where such reactions are of interest to us, the conditions are far from those in the standard state. Concentrations are more likely to be near $10^{-2}\,M$. In this case, for a reaction in which ATP at $10^{-2}\,M$ is hydrolyzed to ADP and P_i at the same concentration, we obtain

$$\Delta G \simeq -10 \text{ kcal/mole} \tag{3.9}$$

What this means, of course, is that ATP hydrolysis is promoted at high dilutions.

In the above, we have been calculating the free-energy change accompanying a given reaction involving given initial and final conditions. Suppose, instead, that we ask that the reaction be carried out reversibly. This means that equilibrium concentrations are maintained. In practice, for a finite amount of reaction to be carried out in this way, it would be necessary to add reactants and take away products so as to maintain the unique set of equilibrium concentrations. In this case, at constant T and P, the total free-energy change for any amount of reaction will be zero. Therefore, Equation (3.6) becomes

$$0 = \Delta G^0 + RT\ln\left(\frac{c_G^g c_H^h \cdots}{c_A^a c_B^b \cdots}\right)_{eq} \tag{3.10}$$

for ideal solutions, and Equation (3.5) yields

$$0 = \Delta G^0 + RT\ln\left[\frac{(c_A y_G)^g (c_H y_H)^h \cdots}{(c_A y_A)^a (c_B y_B)^b \cdots}\right]_{eq} \tag{3.11}$$

for the more general case. Since the concentrations involved must now be the equilibrium concentrations, we have the result

$$\Delta G^0 = -RT\ln\left[\frac{(c_G y_G)^g (c_H y_H)^h \cdots}{(c_A y_A)^a (c_B y_B)^b \cdots}\right]_{eq} = -RT\ln K_t \tag{3.12}$$

This means that the ΔG^0 value will tell us the equilibrium constant for a reaction. This will, in general, be the thermodynamic equilibrium constant, which involves the activity coefficient correction. It can, for dilute solutions, be approximated by the practical equilibrium constant, defined as

$$K = \left(\frac{c_G^g c_H^h \cdots}{c_A^a c_B^b \cdots}\right)_{eq} \tag{3.13}$$

Students are frequently confused at this point. Why is it that the *standard-state* free-energy change can be used to measure the *equilibrium* constant, while the free-energy change at equilibrium is zero? It helps to visualize Equation

---

†Is this always true *in vivo*? It is usually assumed that the in-vitro conditions used to study biochemical reactions correspond roughly to those in the cell. But we really know little about such questions as the concentrations of reactants (including water!) in compartmentalized regions of the cell.

(3.10) or (3.11) as stating something like this: We can pick an arbitrary reference state, the standard state. In general, ΔG will not be zero if reactants and products are in standard states. There will be some set of concentrations (the equilibrium set) at which ΔG will be zero. The second term in Equation (3.10) or (3.11) represents the free-energy change involved in bringing reactants and products from the standard-state concentrations to the equilibrium-state concentrations. This term must just balance ΔG^0 for the overall ΔG to be zero.

For the purpose of interpreting values, it is worthwhile to see how the equilibrium constant varies with ΔG^0: We can rewrite Equation (3.12) as

$$K_t = e^{-\Delta G^0/RT} \tag{3.14}$$

Since $RT \simeq 0.6$ kcal near room temperature, a value of $\Delta G^0 = -0.6$ kcal corresponds to $e^1 = 2.718$, and a value of $\Delta G^0 = -1.2$ kcal corresponds to $K = e^2$, about 8.7. Thus even what seem to be small ΔG^0 values correspond to fairly large values of the equilibrium constant. If we consider an "essentially irreversible" reaction to be one in which $K > 10^4$, this will correspond to a free-energy change more negative than about 5.5 kcal. In other words, such reactions as the hydrolysis of ATP are essentially irreversible.

Another way of measuring the free-energy change in a chemical reaction is through the electrical potential of a cell (real or hypothetical) in which the reaction occurs. It is always possible to calculate the electromagnetic force corresponding to a given reaction, since the free-energy change determines the maximum work other than PV work that the reaction can produce (see Figure 1.4). If we interpret this "other" work as electrical, we have

$$\Delta G = -w_{rev} = -n\mathfrak{F}\varepsilon \tag{3.15}$$

where n is the number of moles of electrons transferred per mole of reaction, and $\mathfrak{F}$ is the Faraday constant (23.06 kcal/V·equiv). This leads to the Nernst equation

$$\varepsilon = \varepsilon^0 - \frac{RT}{n\mathfrak{F}} \ln \frac{(c_G y_G)^g (c_H y_H)^h \cdots}{(c_A y_A)^a (c_B y_B)^b \cdots} \tag{3.16}$$

where ε^0 is the standard-state potential. A table of standard potentials is equivalent to a table of standard free-energy changes, since

$$\Delta G^0 = -n\mathfrak{F}\varepsilon^0 \tag{3.17}$$

and a positive value for the standard potential for a process indicates spontaneity under standard conditions. Biological oxidation-reduction reactions are often thought of in this way.

For many reasons, it is often important to be able to determine the thermodynamic parameters (ΔG^0, ΔH^0, ΔS^0) involved in biochemical processes. Thus a few words about the experimental problems are in order. If it is possible to experimentally determine the equilibrium constant for a reaction, by measuring the concentrations of reactants and products present at equilibrium, Equation (3.12) yields ΔG^0 directly. Two points should be noted: First, in most biochemical systems, activity coefficients are not known; thus K_t in Equation (3.12) must be approximated by the practical equilibrium constant [Equation (3.13)]. Second, the meaning of the value of ΔG^0 obtained will depend on the way in which the reaction is written and the way in which the equilibrium constant is consequently expressed. To take an example: If we express a dimerization reaction by the equation

$$2A \; \rightleftharpoons \; B \tag{3.18}$$

and consequently write

$$K = \frac{c_B}{c_A^2} \tag{3.19}$$

the ΔG^0 obtained will be for the process as written in Equation (3.18): Two moles of A (in standard state) forming 1 mole of B. On the other hand, we could equally well write

$$A \; \rightleftharpoons \; \tfrac{1}{2}B \tag{3.20}$$

and

$$K' = \frac{c_B^{1/2}}{c_A} \tag{3.21}$$

Then the ΔG^0 calculated from K' will be the free-energy change corresponding to the conversion of 1 mole of A into $\frac{1}{2}$ mole of B. It will, of course, be half as large a number as given from Equation (3.19). This is all eminently sensible, but it sometimes causes confusion.

To obtain the enthalpy change corresponding to a given reaction, two courses are available. The most direct way is simply to let the reaction occur in a calorimeter and measure the heat evolved or absorbed. One must, of course, know the number of moles that actually react, and the reaction must usually be fairly rapid to allow precise measurement; it is difficult to measure a small amount of heat that is slowly evolved. To be most precise, of course, one should correct to the standard states of reactants and products (to obtain ΔH^0) by taking into account the heats of dilution of products and reactants. In practice, such corrections are often neglected.

For reactions that are simply not amenable to calorimetric study, more indirect ways of determining ΔH^0 are available. For example, Equation (3.12) can be rewritten as

$$-RT \ln K = \Delta G^0 = \Delta H^0 - T \Delta S^0 \tag{3.22}$$

or

$$\ln K = \frac{-\Delta H^0}{RT} + \frac{\Delta S^0}{R} \tag{3.23}$$

(Note that we have made the practical approximation that $K_t \simeq K$; this will be used henceforth.) Equation (3.23) states that $\ln K$ should be a linear function of $1/T$ if ΔH^0 and ΔS^0 are independent of T. Many times this assumption is valid; often it is not but is made anyway, with unfortunate results. The slope of the straight line obtained by graphing ln K versus $1/T$ (called a van't Hoff graph) should be $-\Delta H^0/R$.

It must be emphasized that this technique, as it is often employed, is beset with traps. Often ΔH^0 varies with T; this will happen, for example, whenever there is a difference in heat capacity between reactants and products. The occurrence of the *logarithm* of K in Equation (3.23) means that any physical quantities *proportional* to the concentrations of reactants and products can be used to define an apparent K, which will still give the right slope. But a hazard exists if these quantities do not measure what the experimenter thinks they measure or if the reaction is a complex one. One may literally measure a "standard-state free-energy change" for a completely meaningless reaction.

3.2 Binding of Small Ligands by Macromolecules

A major biological function of many biopolymers is the binding of small molecules and/or ions. Examples include enzymes, which bind substrates and effector molecules, transport proteins such as hemoglobin, or storage proteins such as myglobin, both of which bind oxygen, and the many proteins that act as buffers by binding hydrogen ions. In fact, almost all biological functions involve the interactions of those small molecules which serve as metabolites, regulators, and signals with the specific surfaces of the macromolecules that carry out cellular processes. For this reason, an understanding of the mechanisms of such interactions is essential to a comprehension of biochemistry at the molecular level.

In most cases, such binding involves the formation of some kind of noncovalent bond between the small molecule or ion (called the *ligand*) and some specific region on or near the surface of the macromolecule. This region will be called the *binding site*. Most biological macromolecules possess binding sites of varying degrees of strength and specificity for a variety of ligands. Since the act of binding a particular ligand may itself induce conformational changes in the biopolymer, which may in turn modify other sites, we may expect that in general the binding of several ligands to one macromolecule will be a complex process. It can be stated, in fact, that it is this very possibility of interdependence of binding affinities that provides a *raison d'être* for the complexity of many biopolymers, for it is in this way that one kind of metabolite can sense the concentration of another kind, even though direct interaction between them is imperceptible.

Binding of small molecules or ions to macromolecules can be so strong as to appear almost irreversible under ordinary conditions (as, for example, the binding of heme to hemoglobin), but unless covalent bonds are involved, we can in principle always consider it to be an equilibrium process. In fact, in most cases the binding will be found to be of such strength that an appreciable concentration of free ligand will be in equilibrium with bound ligand under experimentally accessible conditions.

In investigating such phenomena, the biochemist is usually concerned with the following questions:

1. What is the maximum number of moles of ligand that can be bound per mole of the macromolecule? That is, what is the number of sites, n, for a ligand, designated A?

2. What are the equilibrium constants (and other thermodynamic parameters) for binding of this ligand to each of the sites?

3. Is the equilibrium constant for binding of ligand A to each of these sites independent of whether or not any of the remaining sites for A are occupied? Obviously, this is directly related to question 2.

4. Are the equilibrium constants for binding of ligand A modified by the binding of some other ligand, B, to the same macromolecule?

To show how such questions may be investigated, we shall first describe some of the techniques used in binding studies, concentrating on the question: What are the experimentally available data?

EXPERIMENTAL MEASUREMENTS OF BINDING

In a sense, all measurements of equilibrium binding processes are indirect, in that we cannot observe directly which macromolecules have ligands bound to which binding sites.† In most cases, all we can measure is the fraction of all the ligand molecules in the system that are bound, or the fraction of binding sites that are occupied. The general techniques for binding studies can be divided into classes which yield one or the other of these kinds of information.

An example of the first type is provided by equilibrium dialysis, a technique widely employed for binding studies. This method was described in Section 2.3, where it was shown that, with certain assumptions, one could determine the concentrations of both free and bound ligands in equilibrium with the macromolecules. If we write for the molar concentrations of free ligand and bound ligand [A] and $[A_b]$, respectively, the total ligand concentration $[A_T]$ is obviously

$$[A_T] = [A] + [A_b] \qquad (3.24)$$

If now we designate by $[P_T]$ the total molar concentration of macromolecules, we can calculate the number of moles of A bound per mole of P under the conditions of the particular experiment. This must be an average number since at any instant the dynamic equilibrium between free and bound states will result in different molecules of P having different numbers of sites occupied. We define

---

†This is not strictly true. We can now observe the binding of large molecules to other large molecules in the electron microscope. However, it is very difficult to decide if such observations represent equilibrium binding.

this average number as $\bar{\nu}$:

$$\bar{\nu} = \frac{[A_b]}{[P_T]} \tag{3.25}$$

If we can continue the experiments to high enough concentrations of A that the binding approaches saturation, $\bar{\nu}$ will approach the limit n, the total number of sites per macromolecule available to A. Thus a generalized binding curve will always behave in the manner shown in Figure 3.1; it will be a monotonic function of the free ligand concentration, approaching a finite limit as $[A] \to \infty$.

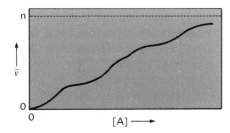

Figure 3.1 A schematic graph of $\bar{\nu}$ versus [A] in some complicated binding process. Two points are general: $\bar{\nu} \to 0$ as $[A] \to 0$, and $\bar{\nu} \to n$ as $[A] \to \infty$.

Unfortunately, the limit (and hence the number n) is often difficult to deduce from the kind of graph shown in Figure 3.1. We shall show later that for certain simplified types of binding, other graphical presentations are more useful.

The equilibrium dialysis technique, although it can yield accurate and complete data, is laborious and demands in many cases large quantities of the substances involved. Variants of this method involving gel filtration have been devised. These are sometimes more convenient, but to date have not proved as accurate.

A quite different kind of technique that is frequently employed in binding studies involves the detection of some physical change in either the macromolecule or the ligand upon binding. An example is the use of light absorption to follow the binding of oxygen by hemoglobin; the spectrum of oxyhemoglobin is different from the spectrum of deoxyhemoglobin. If it can be demonstrated that the change in some physical property (ΔX) with binding is linear in the extent of binding, we may say that

$$\frac{\Delta X}{(\Delta X)_T} = \frac{\bar{\nu}}{n} = \theta \tag{3.26}$$

where $(\Delta X)_T$ is the total change produced when the macromolecules are saturated with A. The quantity θ is often called the *fraction saturation*. Although such techniques may make use of a wide variety of physical measurements, including light absorption, fluorescence, nuclear magnetic resonance, and so forth, there are certain limitations and disadvantages. These include:

1. The change must be linear in $\bar{\nu}$, and be the same for each site in a multi-site molecule (or the precise relationship between change and $\bar{\nu}$ must be known).

It often happens that binding to different sites on a single macromolecule may yield different changes in the property measured. Sometimes only a subset of the different sites for a given ligand will yield a detectable change. Of course, when combined with other studies (such as equilibrium dialysis), this behavior may allow one to distinguish between kinds of sites.

2. In most instances n cannot be determined in this way, for one only determines the *fraction* of total sites which are occupied. A special case occurs, however, if the binding is very strong, for in such cases a binding curve graphed as ΔX versus the *total* concentration of A in the system ($[A_T]$) will be of the form shown in Figure 3.2. Here ΔX has been graphed versus $[A_T]/[P_T]$. If the binding

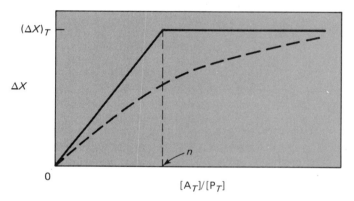

Figure 3.2 The change in some measurable parameter X graphed against *total* ligand concentration for very strong binding (solid line) and weaker binding (dashed curve). In the strong-binding case a break point will be observed at saturation.

is very strong, a curve like the solid line will be obtained; each A molecule added to the system will be bound, changing ΔX, until the system is saturated. Then, even though more A molecules are added, they will not bind, and no more change will be observed. The "break point" of the curve occurs when $[A_T]/[P_T] = n$. Note, however, that if the binding is not strong, a curve such as that shown by the dashed line is observed; it would be very hazardous to attempt to deduce n from this curve.

However, if the value of n is known, considerable information can be extracted from a curve like the dashed line in Figure 3.2. Since $\Delta X/(\Delta X)_T = \bar{\nu}/n$, we can obtain $\bar{\nu}$. Since $\bar{\nu}$ is the average number of moles of ligands bound per mole of P, we can obtain the concentration of bound A by: $[A_b] = \bar{\nu}[P_T]$. Then the concentration of free A will be given by Equation (3.24). This means that we know $\bar{\nu}$ as a function of $[A]$, and as we shall show in the following section, it is this kind of information that can be analyzed to yield binding constants and details of the mechanism of binding.

A final note: Observe that if the binding is very strong (solid line, Figure 3.2), almost *all* A is bound until saturation is reached. This means that the concentration of free A is an indeterminable small quantity, and that further analysis (to yield binding constants, and so forth) will be impossible.

With these preliminary remarks and definitions, we can now see how different kinds of binding, of increasing degrees of complexity, can be analyzed.

THE SIMPLEST CASE:
A SINGLE SITE PER MACROMOLECULE

In order to introduce some of the formalism and to provide a basis for discussion of experimental methods, we shall begin with the particularly simple case in which a macromolecule (P) has only a single site for binding of a ligand A. Furthermore, we shall neglect for the moment the effects of any other ligands. The reaction is then

$$P + A \; \rightleftharpoons \; PA \tag{3.27}$$

The binding constant is then defined as

$$K = \frac{[PA]}{[P][A]} \tag{3.28}$$

where the square brackets denote, as usual, molar concentrations. Remember that [A] in Equation (3.28) must be the concentration of *free* ligand in the mixture.

Sometimes it is more useful to think of the reaction as a *dissociation* process rather than as a binding process:

$$PA \; \rightleftharpoons \; P + A \tag{3.29}$$

in which case we can write the *dissociation constant*

$$\mathbf{K} = \frac{[P][A]}{[PA]} \tag{3.30}$$

Obviously,

$$\mathbf{K} = \frac{1}{K} \tag{3.31}$$

We now wish to express results in terms of the experimentally accessible parameter, $\bar{\nu}$, defined as the average number of bound ligand molecules per macromolecule. In the present case, since the number of moles of bound ligand per unit volume is equal to [PA], and the total number of moles of the macromolecule per unit volume is given by the sum ([P] + [PA]), we have

$$\bar{\nu} = \frac{[PA]}{[P] + [PA]} \tag{3.32}$$

This may be rewritten in a more useful form by using Equation (3.28), in the form $[PA] = K[P][A]$:

$$\bar{v} = \frac{K[P][A]}{[P] + K[P][A]}$$

$$\bar{v} = \frac{K[A]}{1 + K[A]} \tag{3.33}$$

Equation (3.33) describes $\bar{v}$ as a function of the free ligand concentration [A]. The graph of $\bar{v}$ versus [A] is a rectangular hyperbola like the curve in Figure 3.3.

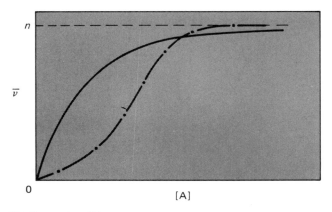

Figure 3.3 Curves describing noncooperative binding (solid line) and cooperative binding (dashed line). The former follows Equation (3.49), or with $n = 1$, Equation (3.33). The latter, which has a sigmoidal shape, might be described by an equation such as (3.64).

It is clear from Equation (3.33) that $\bar{v} \rightarrow 1$ (saturation) only as $[A] \rightarrow \infty$. At half-saturation ($\bar{v} = 0.5$), $[A] = 1/K$. Note that if we had written Equation (3.33) in terms of a dissociation constant [Equation (3.31)] we would have (show this)

$$\bar{v} = \frac{[A]/K}{1 + [A]/K} \tag{3.34}$$

and $[A] = K$ at $\bar{v} = 0.5$. This points up one convenience of dissociation constants: at least in such simple cases, the numerical value of the dissociation constant (in moles per liter) shows immediately the ligand concentration at which the system is half-saturated.

It is not easy to decide by graphing binding data according to Equation (3.33) whether this simple mechanism of binding applies in any particular case. For such purposes, it is better to rearrange Equation (3.33) to produce a function that yields a linear graph. There are several ways of doing this. One formulation, known as a *Scatchard plot*, is widely used.

$$\frac{\bar{\nu}}{[A]} = K - K\bar{\nu} \qquad (3.35)$$

A graph of $\bar{\nu}/[A]$ versus $\bar{\nu}$ should be linear. As we shall see later, the Scatchard plot is particularly useful in determining the stoichiometry of binding when more than one ligand is bound per macromolecule.

Another linear variant of Equation (3.33) will turn out to have much wider application. This is obtained by forming the ratio $\bar{\nu}/(1 - \bar{\nu})$, which yields (show this)

$$\frac{\bar{\nu}}{1 - \bar{\nu}} = K[A] \qquad (3.36)$$

This is a special form of the Hill equation (see below). If one graphs $\log [\bar{\nu}/(1 - \bar{\nu})]$ versus $\log [A]$, a straight line of unit slope should be obtained if Equation (3.33) is obeyed.† The value of A at $\bar{\nu}/(1 - \bar{\nu}) = 1$ will equal $1/K$.

In this section we have made the restrictive assumption that $n = 1$. Although there are many cases in which a macromolecule has but a single binding site for a particular ligand, most of the interesting situations arise when multiple sites are present. We turn now to such examples.

MULTIPLE EQUILIBRIA

In this section we shall consider those cases in which each macromolecule can bind n ligand molecules, n being a number greater than 1. We shall find that in this event, interaction between the binding sites can produce complex and important effects. These include the whole range of "allosteric" effects of such great importance in enzyme chemistry.

Some General Relationships. If there are n sites per molecule, the expression for $\bar{\nu}$ becomes a bit more complicated than in the single-site case. As before, we define $\bar{\nu}$ as the average number of ligands bound per macromolecule. However, there may now be some molecules binding one ligand, some binding two, some binding three, and so forth, up to n. So

concentration of bound ligand $= [PA] + 2[PA_2] + 3[PA_3] + \cdots$

$$= \sum_{i=1}^{n} i[PA_i] \qquad (3.37)$$

The numerical coefficients enter, of course, because a mole of $[PA_i]$ carries i moles of ligand. The total molar concentration of macromolecule is just the sum of its molar concentrations in all forms:

concentration of macromolecule $= \sum_{i=0}^{n} [PA_i] \qquad (3.38)$

---

†The discerning student may wonder why we would use a logarithmic plot, when a linear graph of $\bar{\nu}/(1 - \bar{\nu})$ versus A would suffice. The reason will appear in the next section.

Note that this summation starts at zero, to include the term [P], corresponding to unliganded macromolecule. Now $\bar{v}$ becomes

$$\bar{v} = \frac{\sum\limits_{i=1}^{n} i[PA_i]}{\sum\limits_{i=0}^{n} [PA_i]} \tag{3.39}$$

This statement is completely general; it includes Equation (3.32) as the special case where $n = 1$.

In order to proceed further with analysis, we must have some relationship between the $[PA_i]$. This can be done in a formal sense by writing a series of equilibrium equations. A number of formulations are possible; two are shown below.

Formulation I   Formulation II

$$P + A \rightleftharpoons PA \quad k_1 = \frac{[PA]}{[P][A]} \qquad P + A \rightleftharpoons PA \quad K_1 = \frac{[PA]}{[P][A]}$$

$$PA + A \rightleftharpoons PA_2 \quad k_2 = \frac{[PA_2]}{[PA][A]} \qquad P + 2A \rightleftharpoons PA_2 \quad K_2 = \frac{[PA_2]}{[P][A]^2}$$

$$\vdots \qquad\qquad\qquad\qquad\qquad \vdots$$

$$PA_{n-1} + A \rightleftharpoons PA_n \quad k_n = \frac{[PA_n]}{[PA_{n-1}][A]} \qquad P + nA \rightleftharpoons PA_n \quad K_n = \frac{[PA_n]}{[P][A]^n}$$

$$\tag{3.40}$$

These ways of describing the equilibria are equivalent. One set of equilibrium constants can always be expressed in terms of the other; for example, $K_i = k_1 k_2 k_3 \cdots k_i$. In any event, it should be understood that the manner in which the equilibrium constants are written says nothing about the mechanism of the reaction; for example, writing $K_n = [PA_n]/[P][A]^n$ does *not* imply that we believe n ligands bind simultaneously to a molecule of P. In many cases the second formulation leads to simpler equations, and we shall use it here.

Using the set of equations in formulation II, we may rewrite Equation (3.39) as

$$\bar{v} = \frac{\sum\limits_{i=1}^{n} iK_i[P][A]^i}{\sum\limits_{i=0}^{n} K_i[P][A]^i} \tag{3.41}$$

where K_0 has been defined as equal to unity. Canceling [P], which factors out, we find that

$$\bar{v} = \frac{\sum\limits_{i=1}^{n} iK_i[A]^i}{\sum\limits_{i=0}^{n} K_i[A]^i} \tag{3.42}$$

This very general binding equation is known as the *Adair equation*. Equation (3.33) is the special case where $n = 1$. The Adair equation will, in principle, describe almost any binding situation, for nothing has been assumed about the behavior of the individual K_i's. However, if n is large (even as large as 4), there are so many adjustable parameters in Equation (3.42) that their determination with precision requires extensive and exact binding data. Furthermore, simply to fit data to the Adair equation is not especially revealing about the mechanism of binding. Often, it is more useful to see if the data can be fitted by some more restrictive equation, which embodies a simple model and requires fewer adjustable parameters. We shall turn now to the consideration of some such special cases.

All Sites Equivalent and Independent. The simplest case of a multiple equilibrium occurs when each of the n sites on a biopolymer has the same affinity for the ligand as any other, and the binding is *noncooperative*; that is, the affinity of any site is independent of whether or not other sites are occupied. At first glance, it might seem that this would simply mean that all of the k_i were equal. However, this is *not* the case, and to understand why we must look more closely at the equilibrium constants we have defined above. In writing, for example, $K_i = [PA_i]/[P][A]^i$ or $k_i = [PA_i]/[PA_{i-1}][A]$, the concentrations, such as $[PA_i]$ or $[PA_{i-1}]$, each represent the sum of concentrations of a whole class of molecules. Consider, for example, the set of molecules shown in Figure 3.4, in each of which two ligands are bound to a four-site macromolecule.

Figure 3.4 Four of the six (4!/2!2!) arrangements of two ligands over four sites. Note that the total concentration of PA_2 will be the sum of the concentrations of these six forms.

The number of such isomers of a particular liganded state is given by the number of ways in which n sites may be divided into i occupied sites and $(n - i)$ vacant sites. This is

$$N_{i,n} = \frac{n!}{(n - i)!\, i!} \tag{3.43}$$

In our macroscopic measurements we do not make distinctions between such isomers, and the *macroscopic* equilibrium constants we have written (the K_i or k_i) simply lump all such isomers together. But when we are talking about the *affinity* of a particular site, we are referring to a reaction in which a ligand molecule becomes bound to one particular site. The *microscopic* constant for such a reaction would have to be written

$$(k_i)_{kl} = \frac{[PA_{i,l}]}{[PA_{i-1,k}][A]} \tag{3.44}$$

which refers to the addition of a ligand to a particular isomer (k) of the class $[PA_{i-1}]$ at a specific vacant site to form a particular member (l) of the class $[PA_i]$.

Since the number of ways in which this addition can be made depends on the value of i, it turns out that even in the case of interest here, where all the microscopic constants are equal, the macroscopic constants are not equal. In order to evaluate the individual terms in Equation (3.39) for this case, we must consider first the formation of a particular species such as $[PA_i]_l$. We can imagine it to be made by the addition of successive ligands to P. Since the microscopic association constants k_j for each step have been assumed to be equal, we give their value as k, and write

$$[PA_i]_l = k^i[P][A]^i \tag{3.45}$$

which will be valid whatever the path of additions we choose. All of the individual species of $[PA_i]$ will have the same (average) concentration. Therefore, for the total concentration of $[PA_i]$ we need only multiply Equation (3.45) by the statistical factors $N_{i,n}$:

$$[PA_i] = \frac{n!}{(n-i)!\,i!}[P]k^i[A]^i \tag{3.46}$$

This makes the Adair equation take the somewhat formidable form

$$\bar{v} = \frac{\sum\limits_{i=1}^{n} i\,\dfrac{n!}{(n-i)!\,i!}\,(k[A])^i}{\sum\limits_{i=0}^{n} \dfrac{n!}{(n-i)!\,i!}\,(k[A])^i} \tag{3.47}$$

However, Equation (3.47) is easily simplified, for the sums in the numerator and denominator are closely related to the binomial expansion. In fact, the denominator is simply the binomial expansion of $(1 + k[A])^n$. The numerator may be evaluated by noting that

$$(1 + k[A])^{n-1} = \sum_{i=0}^{n-1} \frac{(n-1)!}{(n-1-i)!\,i!}k^i[A]^i$$

$$= \frac{1}{nk[A]}\sum_{i=0}^{n-1} \frac{n!\,(i+1)}{(n-(i+1))!\,(i+1)!}k^{i+1}[A]^{i+1} \tag{3.48}$$

But the sum in Equation (3.48) is exactly the sum in the numerator of Equation (3.47) if we replace $i + 1$ by a new index $j = i + 1$. (It does not matter what symbol we give for the index.) Therefore,

$$\bar{v} = \frac{nk[A](1 + k[A])^{n-1}}{(1 + k[A])^n}$$

$$= \frac{nk[A]}{1 + k[A]} \tag{3.49}$$

Equation (3.49) looks remarkably like the equation for single-site binding [Equation (3.33)], differing only in the inclusion of n. The similarity is not accidental, and it suggests a very simple way of thinking about Equation (3.49). If all of the n sites on the macromolecule are indeed equivalent and independent of one another, there is no way in which a ligand molecule, binding to such a site, can "know" that it is binding to a site that is somehow attached to other sites. The fraction of all the sites in the system that are occupied is $\bar{v}/n$ (since there are $\bar{v}$ sites occupied per molecule, out of a total of n sites per molecule). If the sites are independent, the fraction occupied should simply follow Equation (3.33):

$$\frac{\bar{v}}{n} = \frac{k[A]}{1 + k[A]} \tag{3.50}$$

which is identical to Equation (3.49).

The similarity of Equation (3.49) to Equation (3.33) means that the variants of Equation (3.33) we considered can easily be rewritten for this simple case of multiple binding. A particularly widely used form is the Scatchard equation, analogous to Equation (3.35):

$$\frac{\bar{v}}{[A]} = k(n - \bar{v}) \tag{3.51}$$

Extrapolation of a graph of $\bar{v}/[A]$ versus $\bar{v}$ will yield the value $\bar{v} = n$ on the $\bar{v}$ axis, where $\bar{v}/[A] = 0$ (see Figure 3.5). The method is widely used, but must be em-

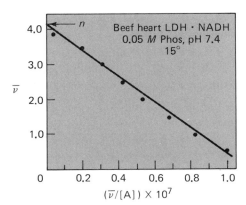

Figure 3.5 A graph according to Equation (3.51). This illustrates the binding of NADH to beef heart LDH. Evidently, the binding is noncooperative, and $n = 4$. Data from S. Anderson and G. Weber, *Biochemistry.* **4**, 1948 (1965). Reprinted by permission of the American Chemical Society.

ployed with caution, for the extrapolation to the limit where $\bar{v}/[A] \to 0$ can be somewhat uncertain. This will be particularly true in the cases where more than one type of site is present, as will be discussed in the next section.

Binding to Nonequivalent Sites. Considering the complexity of the structure of most proteins and other biopolymers, it is not surprising that in many cases a number of different *kinds* of sites for a given ligand may be found on one macromolecule. In this event, we should expect, a priori, that these different kinds of sites will have different binding constants for the ligand. It should be emphasized that we are *not* considering cooperative binding here; the value of the association constant for each site is still assumed to be independent of the state of occupancy of all other sites. Very often there is a large difference in affinity between the strongest sites and the weakest; and one may have to study the binding over a very wide range in ligand concentration to see the whole picture.

The formal analysis of such binding situations is a straightforward extension of what we have considered so far. Suppose that there are N classes of sites, numbered $S = 1 \rightarrow N$, and that there are n_S sites in each class (n_S can, of course, be as small as 1). Then $\bar{\nu}$, the total number of sites occupied by a ligand per macromolecule, will simply be the sum of the $\bar{\nu}_S$ values, the numbers of sites of each class occupied per macromolecule.

$$\bar{\nu} = \sum_{S=1}^{N} \bar{\nu}_S \tag{3.52}$$

If each class of sites binds independently, with an intrinsic association constant k_S, we have, by Equation (3.49),

$$\bar{\nu}_S = \frac{n_S k_S [A]}{1 + k_S [A]} \tag{3.53}$$

so

$$\bar{\nu} = \sum_{S=1}^{N} \frac{n_S k_S [A]}{1 + k_S [A]} \tag{3.54}$$

To investigate the behavior of such systems, let us consider a simple case in which there are only two classes of sites: n_1 "strong" sites each with a microscopic association constant k_1, and n_2 "weak" sites with constant k_2. By definition, $k_1 > k_2$. Then Equation (3.54) reduces to

$$\bar{\nu} = \frac{n_1 k_1 [A]}{1 + k_1 [A]} + \frac{n_2 k_2 [A]}{1 + k_2 [A]} \tag{3.55}$$

Unless $k_1 \gg k_2$, it is very difficult to distinguish this situation from binding by $(n_1 + n_2)$ equivalent sites with k intermediate between k_1 and k_2 (see Figure 3.6

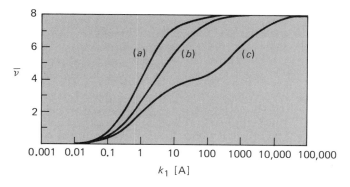

Figure 3.6 Examples of the kinds of binding curves obtained if there are two classes of sites. It is presumed that there are eight sites in all, with four in each class 1 and 2. Curve (*a*) is obtained if $k_1 = k_2$, curve (*b*) if $k_1 = 10k_2$, and curve (*c*) if $k_1 = 1000k_2$. In the case of (*b*) it is difficult to tell that there are two classes.

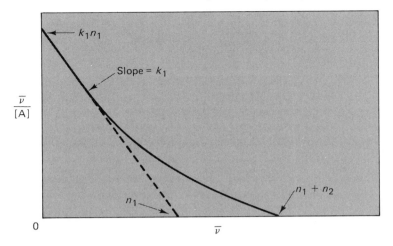

Figure 3.7 A graph of the kind of Scatchard plot that might be observed for binding to a molecule with two types of sites; there are n_1 strong sites and n_2 weak sites.

for examples). In such cases Scatchard plots are often useful, for such a graph for Equation (3.55) will appear as in Figure 3.7. If $k_1 \gg k_2$, the initial binding (low $\bar{v}$) will be dominated by k_1, so the first part of the curve will be given approximately by

$$\bar{v} \simeq \frac{n_1 k_1 [A]}{1 + k_1 [A]} \tag{3.56}$$

Thus extrapolation of this region to the abscissa should give n_1, and the initial slope will yield k_1 (see Figure 3.7). On the other hand, the intercept of the full curve with the abscissa will yield $(n_1 + n_2)$, allowing calculation of n_2, and by more indirect means, k_2. Note, however, that for the method to be really safe, k_1 and k_2 must be quite different. Furthermore, if there are several classes of sites the analysis becomes very uncertain. In principle, the intercept with the abscissa should always give the total number of sites, but in many cases the curve approaches the $\bar{v}$ axis almost asymptotically, and determination of the intercept is very uncertain.

The existence of different classes of sites can also be demonstrated by the use of Hill plots. One graphs $\log [\bar{v}/(n - \bar{v})]$ (or $\log [\theta/(1 - \theta)]$, where $\theta = \bar{v}/n$) versus $\log [A]$ (Figure 3.8). The data obtained at the lowest values of $[A]$ lie close to the line corresponding to the strong binding sites. As $[A]$ increases, the data move over to approach the line corresponding to the weakest class of sites. Both k_1 and k_2 can be *estimated*, but n_1 and n_2 are not given by this method. Obviously, in such cases a combination of two or more graphical techniques may be very useful.

Cooperative Binding. If binding of a ligand to one site on a macromolecule influences the affinity of other sites, the binding is said to be *cooperative*. Such

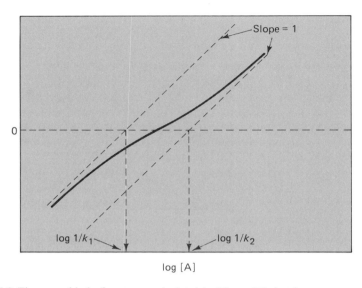

log [A]

Figure 3.8 The same kind of system as depicted in Figure 3.7, but here represented by a Hill plot. Here $n = n_1 + n_2$. The ordinate could equally well be written as $\log [\theta/(1 - \theta)]$, where $\theta = \bar{v}/n$ is the fraction saturation. A similar curve would be observed for negative cooperativity.

cooperativity can be *positive* (binding at one site increases the affinity of others) or *negative* (if the affinity of other sites is decreased). Such effects are also referred to as *allostery*, a general term that includes *homeoallostery*, which refers to the influence on ligand binding by ligands of the same kind, and *heteroallostery*, in which binding of ligands of one kind modifies the binding of a second kind of ligand.

For the present, we shall be concerned with homeoallosteric effects. Before proceeding to a detailed analysis of some of the models that have been proposed to explain such behavior, let us examine it in a qualitative way.

The differences in binding curves between a noncooperative and a positively cooperative binding are illustrated in Figure 3.3. The noncooperative curve is that given by Equation (3.49); it is a rectangular hyperbola. The cooperative curve exhibits a *sigmoid* shape. This can be qualitatively explained in the following way: At first binding is very weak; the molecule without ligands is presumed to be in a weakly binding state, and the first ligands bound are mostly going to different molecules. But filling the first site on any molecule somehow increases the affinity of the remaining sites, and binding becomes stronger as more sites are filled. So the curve turns upward, giving it the sigmoidal shape.

The effect of positive cooperativity is seen even more clearly on a Hill plot (Figure 3.9). The points corresponding to binding of the first ligands in the system fall near the "weak binding state" line. As more sites are filled, the points move over toward a line corresponding to the "strong binding state." As

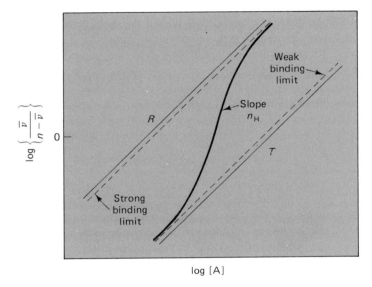

log [A]

Figure 3.9 A Hill plot for positively cooperative binding, interpreted in terms of the MWC theory. The solid lines R and T, which have slopes of unity, correspond to pure R and T states, respectively. The dashed lines are given by Equations (3.67) and (3.66).

the last few openings are filled, almost all molecules will be in this state, and the points will tend toward a line characteristic of this state.

Negative cooperativity would have the opposite effect, and a Hill plot would look very much as shown in Figure 3.8 for a system with fixed strong and weak sites. Note that in all cases the curves will approach, at both low and high values of log [A], straight lines which have slopes of unity. If *positive* cooperativity occurs, the curve must have a slope greater than unity near the middle of the binding range; if *negative* cooperativity is found, there will be a range where the slope is less than unity. Thus Hill plots can be useful as diagnostic tests for binding type:

1. If a straight line with slope of unity is found over the whole range, binding is noncooperative and the sites are identical.

2. If the curve has in some region a slope greater than unity, the binding must be positively cooperative.

3. If the curve has in some region a slope less than unity, the macromolecule either has more than one class of sites, or the binding is negatively cooperative.

It is of interest to consider one idealized case which, while never really encountered, illustrates an extreme of behavior. Suppose that a system were so strongly positively cooperative that each molecule, on accepting one ligand, would be so activated that it would immediately fill the remainder of its n sites before sites on any other molecules were filled. In this "all or none" case, there will be only two macromolecular species present in appreciable quantity: P and PA$_n$. Then Equation (3.39) would reduce to the simple form

$$\bar{\nu} = \frac{n[\text{PA}_n]}{[\text{P}] + [\text{PA}_n]} \tag{3.57}$$

and Equation (3.42) would become

$$\bar{\nu} = \frac{nK_n[\text{A}]^n}{1 + K_n[\text{A}]^n} \tag{3.58}$$

The Hill equation then takes the simple form

$$\log \frac{\bar{\nu}}{n - \bar{\nu}} = \log k_n + n \log [\text{A}] \tag{3.59}$$

which will yield a straight line with slope n. Since this unreal case represents the maximum possible cooperativity, the greatest slope a Hill plot can ever have is n, the number of sites on the macromolecule. In real systems the maximum Hill slope, n_H, will always be less than this, so we can say that for a positively co-operative system

$$1 < n_H < n \tag{3.60}$$

The closer the quantity n_H approaches to n, the stronger is the cooperativity.

In order to say anything more quantitative about cooperative binding, a model is required. There have been two principal kinds of models proposed. We shall now consider these.

The theory developed in 1965 by Monod, Wyman, and Changeux (MWC), and all variants of it, are based on the concept of *concerted* conformational transitions in the subunits of a multisite protein. It is assumed that each subunit can exist in two states; in one state the binding site is weak, and in the other the binding site is strong. The two states have been designated as T and R, respectively. A basic assumption of the MWC theory is that the symmetry of the molecule is maintained; that is, all subunits must undergo the T $\longrightarrow$ R transition in concert. An equilibrium is presumed to exist between these forms, as depicted in Figure 3.10(a), so that at any moment a mixture of these two states will exist. In the absence of ligand, this equilibrium is described by the constant L:

$$L = \frac{[\text{T}]}{[\text{R}]} \tag{3.61}$$

L is assumed to be a large number, so that in the absence of binding most mole-cules are in the T state. The basic idea is that binding of ligands, which occurs more strongly to the R state, will shift this equilibrium so as to favor the R form, in accord with Le Châtelier's principle. This provides a simple mechanism for cooperativity, for adding ligand to the system increases the proportion of unoccupied sites in the R conformation, which then will bind with greater affinity. The R and T forms of the molecule will each, by virtue of the symmetry assumption, be a state with n identical sites. We designate the microscopic bind-

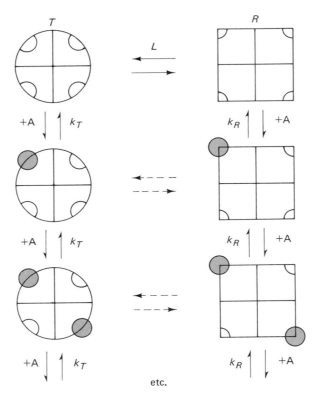

etc.

Figure 3.10 Model for cooperative binding, using a tetramer. (*a*) The Monod–Wyman–Changeux model: The molecule is assumed to exist in equilibrium between two conformational states, T and R. The equilibria between liganded and unliganded forms are described in terms of two microscopic binding constants, k_T and k_R, where $k_R > k_T$. An equilibrium constant L connects these forms. Although one could write equations for other equilibria (dashed arrows), these are not necessary to define the system. Cooperativity arises from the fact that the greater ligranding to the R form pulls the equilibria to the right, therefore increasing the concentration of strong sites.

ing constants for the R and T states by k_R and k_T, respectively, and the ratio of these (k_T/k_R) as c. Since T is defined as the weak-binding state, $c < 1$; often it is a quite small number. With these assumptions and the definition of $\bar{v}$, we may write

$$\bar{v} = \frac{\sum_{i=1}^{n} i[\mathrm{RA}_i] + \sum_{i=1}^{n} i[\mathrm{TA}_i]}{\sum_{i=0}^{n} [\mathrm{RA}_i] + \sum_{i=0}^{n} [\mathrm{TA}_i]} \tag{3.62}$$

Using the same reasoning as that used in deriving Equation (3.47), we obtain

$$\bar{v} = \frac{[R]\sum\limits_{i=1}^{n} i \dfrac{n!}{(n-i)!\,i!} k_R^i[A]^i + [T]\sum\limits_{i=1}^{n} i \dfrac{n!}{(n-i)!\,i!} k_T^i[A]^i}{[R]\sum\limits_{i=0}^{n} \dfrac{n!}{(n-i)!\,i!} k_R^i[A]^i + [T]\sum\limits_{i=0}^{n} \dfrac{n!}{(n-i)!\,i!} k_T^i[A]^i} \tag{3.63}$$

If we now insert $[T] = L[R]$, and evaluate the sums as in deriving Equation (3.49), we find that

$$\bar{v} = n k_R[A]\frac{(1 + k_R[A])^{n-1} + Lc(1 + ck_R[A])^{n-1}}{(1 + k_R[A])^{n} + L(1 + ck_R[A])^{n}} \tag{3.64}$$

where we have used $k_T = ck_R$. The result is certainly not simple, but the behavior of systems obeying the MWC model can be visualized if we rewrite Equation (3.64) in the form appropriate for a Hill plot. After some algebra, we find that

$$\frac{\bar{v}}{n - \bar{v}} = k_R[A]\frac{1 + Lc\left(\dfrac{1 + ck_R[A]}{1 + k_R[A]}\right)^{n-1}}{1 + L\left(\dfrac{1 + ck_R[A]}{1 + k_R[A]}\right)^{n-1}} \tag{3.65}$$

In the limits as $[A] \longrightarrow 0$ and $[A] \longrightarrow \infty$ we obtain

$$\frac{\bar{v}}{n - \bar{v}} \longrightarrow k_R[A]\frac{1 + Lc}{1 + L} \qquad \text{as } [A] \longrightarrow 0 \tag{3.66}$$

$$\frac{\bar{v}}{n - \bar{v}} \longrightarrow k_R[A]\frac{1 + Lc^n}{1 + Lc^{n-1}} \qquad \text{as } [A] \longrightarrow \infty \tag{3.67}$$

Each of these limiting forms corresponds to a straight line of slope equal to unity on a Hill plot. Such lines are shown in Figure 3.9. If L is large (the T form highly favored in the absence of ligand), Equation (3.66) is given approximately by

$$\frac{\bar{v}}{n - \bar{v}} \simeq k_R[A]c = k_T[A] \qquad ([A] \longrightarrow 0) \tag{3.68}$$

Thus the system begins, at low $[A]$, by behaving like the T state. If it is also true that c is small, then, especially if n is large, we find from Equation (3.67) that

$$\frac{\bar{v}}{n - \bar{v}} \simeq k_R[A] \qquad ([A] \longrightarrow \infty) \tag{3.69}$$

That is, in this case the high $[A]$ limit is the R state. It should be emphasized that the Hill plot will approach lines of unit slope at very low or very high $[A]$ in any case; only under the conditions given above will these lines correspond to pure T or R states. The Hill plot must always lie between these limiting lines (R and T in Figure 3.9). The slope is a complicated function of $[A]$, but it will always be less than n.

It is evident from Figure 3.9 that the MWC model predicts the kind of behavior observed in homeoallosteric systems. The model has been extended and generalized by others, and can certainly describe a wide range of experimental data. But is it unique? Does it correspond to what really happens? Others have suggested that cooperativity can be explained in other ways. We now consider a quite different, alternative model.

The model proposed in 1966 by Koshland, Nemethy and Filmer (the KNF model) takes a quite different approach. Rather than assume symmetry in the conformational transformation of subunits, it is assumed that subunits may change *one at a time* from a weak-binding form, called A, to a strong-binding form, called B. Thus mixed states of the kinds shown in Figure 3.10(*b*) are to be expected. The cooperativity arises because interaction between different pairs of

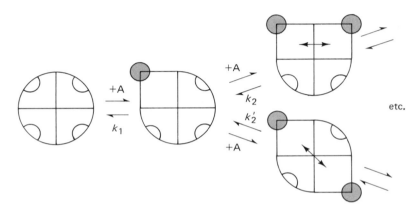

Figure 3.10 Model for cooperative binding, using a tetramer. (*b*) The Koshland–Nemethy–Filmer model: It is assumed that binding of a ligand to a site induces the subunit bearing that site to adopt a new conformation. Interactions with contiguous subunits may then be such as to promote their conversion to the "binding conformation," thus enhancing binding. Different subunit interactions may have different results, leading to considerable complexity in the theory. Note that there exists the possibility that interactions between subunits in the "binding conformation" might be unfavorable; this would yield negative cooperativity.

subunits is assumed to depend on the states of the subunits in the pairs. That is, the stability of pairs A–A, A–B, and B–B is assumed to differ. It is clear that this kind of mechanism is capable of giving rise to a wide range of behavior patterns. One feature that makes the theory of Koshland et al. more complex than that of Monod et al. is that the possible patterns of interaction depend on the geometry of the molecule, or more precisely, on the topology of the possible interactions [see Figure 3.10(*b*)].

In any event the model leads to much more complex expressions for the binding curve than does the MWC model. In terms of fitting individual binding curves, the two models appear to do about equally well with the best data available at the present time. So it appears at this moment that there is little to choose between these models on the basis of binding data alone. In many cases the MWC model is simpler to use; we shall now employ it for a brief discussion of heterotropic allostery.

Heterotropic Effects. In the MWC model, the behavior of the system is strongly dependent on the value of L. If L is very, very large, the macromolecules will tend to remain in the T state over the entire experimentally accessible binding range; if L is small, the molecules will yield a noncooperative binding curve close to that of the R state. Obviously, if some other factor could influence the T $\rightleftharpoons$ R equilibrium described by L, the binding behavior could be profoundly altered. This is how the MWC theory explains heterotropic allostery. An *effector* molecule shifts the value of L. If it makes L greater, it is called an *allosteric inhibitor*; if it makes L smaller, it is an *allosteric activator*. Sometimes it is found that there are a number of molecules or ions which can thus modify the binding behavior for a particular ligand. An example is shown in Figure 3.11.

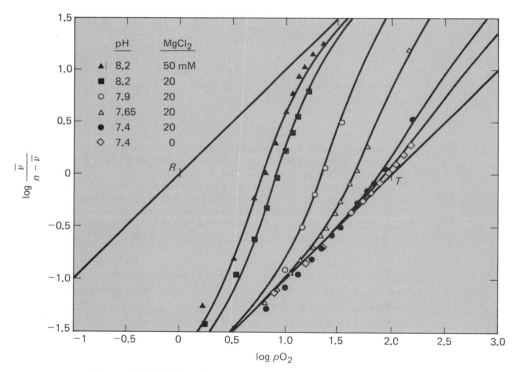

Figure 3.11 Binding of oxygen by the hemocyanin of the ghost shrimp, *Callianassa*. The curves through the points have been fitted to a modification of the MWC theory, using the R and T states shown. From F. Arisaka and K. E. van Holde, *J. Mol. Biol.*, **143**, 41–73 (1979).

3.3 Proton Binding: Titration Curves

The conventional pH titration experiment, whether carried out with a low-molecular-weight substance or with a macromolecule, is a binding study. The ligand (H^+) is quantitatively added to the solution (or removed by adding a strong base) and the concentration of free ligand measured with a sensitive pH electrode. The amount of bound ligand can then be calculated directly. Thus it is somewhat unfortunate that discussions of the process of the binding of protons have traditionally been couched in rather different terms than those used for most binding studies. For while protons behave very much like other ligands, the equations customarily used *look* quite different from those we have been employing here. In the first place, the K_a values usually given are *dissociation* constants. Second, instead of $\bar{v}$, the number of ligands bound, those working with proton–molecule equilibria are prone to speak in terms of $\bar{r}$, the average number of protons dissociated from the fully protonated macromolecule. If the total number of dissociable protons is n, then obviously

$$\bar{r} = n - \bar{v} \tag{3.70}$$

We shall be interested first in the case where there are n independent sites: writing Equation (3.49) in terms of a microscopic dissociation constant, we get

$$\bar{v} = \frac{n[H^+]/\mathbf{k}}{1 + [H^+]/\mathbf{k}} \tag{3.71}$$

so

$$\bar{r} = n - \bar{v} = n\left(1 - \frac{[H^+]/\mathbf{k}}{1 + [H^+]/\mathbf{k}}\right) \tag{3.72}$$

$$= \frac{n\mathbf{k}/[H^+]}{1 + \mathbf{k}/[H^+]} \tag{3.73}$$

It is in this latter form that titration curves are usually expressed. Obviously,

$$\frac{\bar{r}}{n - \bar{r}} = \frac{\alpha}{1 - \alpha} = \frac{\mathbf{k}}{[H^+]} \tag{3.74}$$

where α is the fraction dissociation. So

$$\log \frac{\alpha}{1 - \alpha} = \log \mathbf{k} - \log [H^+] \tag{3.75}$$

$$\log \frac{\alpha}{1 - \alpha} = pH - pK_a \tag{3.76}$$

where $pK_a = -\log \mathbf{k}$. The familiar Henderson–Hasselbalch equation thus turns out to be equivalent to the Hill equation!

In the titration of any real protein, there will not be a single kind of titratable group, but rather a range, from very acidic to very basic. Roughly, these can be grouped into classes—the carboxyls, the imidazoles, the lysines, and so forth. Thus one might attempt to approximate the whole titration curve by an equation analogous to Equation (3.54): that is,

$$\bar{r} = \frac{n_1 \mathbf{k}_1/[H^+]}{1 + \mathbf{k}_1/[H^+]} + \frac{n_2 \mathbf{k}_2/[H^+]}{1 + \mathbf{k}_2/[H^+]} + \cdots \qquad (3.77)$$

A protein titration curve thus looks like Figure 3.12, with overlapping regions for the titration of different kinds of groups. Direct curve fitting to Equation

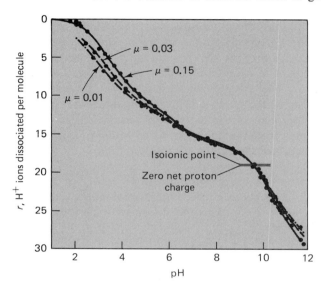

Figure 3.12 The titration of ribonuclease at 25°C. From Tanford and Hauenstein (1956). Reprinted by permission of the American Chemical Society.

(3.77) of curves such as that of Figure 3.12 is rather unprofitable, for the number of adjustable parameters involved is too great. Obviously, more information is needed from other kinds of measurements. More seriously, Equation (3.77) is only a very poor approximation, fundamentally because the affinity of a given group for a proton must be a function not only of the intrinsic properties of the group, but also of that group's environment. The environments for a single kind of group (carboxyl, for example) in a protein molecule may vary widely, from highly hydrophobic regions to surface regions, or to strong interactions with specific other groups (lysine, for example). So a given kind of group may show a range of pK_a values. Furthermore, the environment may change in a number of ways as the titration proceeds.

An obvious change that occurs is in the net charge on the protein molecule. At low pH, the average protein will be positively charged (all the carboxylates will be neutralized, and basic groups such as lysine will be positively charged). As the pH is increased, the protein will go through a pH of zero charge (the isoelectric point) and then, at high pH, adopt a negative charge, since carboxylates will now be anionic, and the basic groups will lose their protons and become neutral.

As this change in charge occurs, the free-energy change involved in removal of a proton from the molecule will change, for the electrical work done against the potential field of the protein will depend on the net charge. We may write the standard-state free-energy change for ionization of some group as

$$\Delta G^0 = \Delta G_{in}^0 + \Delta G_{el}^0 \qquad (3.78)$$

where ΔG_{in}^0 is the "intrinsic" contribution, which depends only upon the intrinsic ionizability of the group, and ΔG_{el}^0 represents the extra work done against the net potential field of the protein.

Now since $\Delta G^0 = -RT \ln \mathbf{k}$, we have

$$-RT \ln \mathbf{k} = -RT \ln \mathbf{k}_{in} + \Delta G_{el}^0 \qquad (3.79)$$

$$\log \mathbf{k} = \log \mathbf{k}_{in} - \frac{\Delta G_{el}^0}{2303RT} \qquad (3.80)$$

or

$$pK_a = pK_{a,in} + \frac{\Delta G_{el}^0}{2.303RT} \qquad (3.81)$$

To a first approximation, we can consider that the electrical term will be given by

$$\frac{\Delta G_{el}^0}{RT} = -2w\bar{z} \qquad (3.82)$$

where w is a positive parameter whose value should depend on the dielectric constant, the ionic strength, and other properties of the medium, and $\bar{z}$ is the average net charge on the protein. The factor 2 is present by convention. Note that if z is $+$, ΔG_{el}^0 is negative; a positive charge on the protein molecule favors the dissociation of protons. Conversely, if $\bar{z}$ is negative, ΔG_{el}^0 is positive; protons are held more strongly by a negatively charged molecule. Using Equation (3.82), we may now write Equation (3.81) as

$$pK_a = pK_{a,in} - 0.868w\bar{z} \qquad (3.83)$$

Thus Equation (3.76) may be rewritten as

$$pH - \log \frac{\alpha}{1 - \alpha} = pK_{a,in} - 0.868w\bar{z} \qquad (3.84)$$

This result provides a method for evaluating the intrinsic pK from protein titration curves. If one can follow the titration of a single group or class of groups (see below), and evaluate both α and net charge $\bar{z}$ at different pH values, a graph of the left side of Equation (3.84) versus z will yield $pK_{a,in}$ as a limit where $\bar{z} = 0$ (see Figure 3.13).

In recent years, it has become possible to know the positions of individual charged groups on a protein molecule quite exactly as a consequence of X-ray diffraction analysis (see Chapter 11). With such information at hand, a much more sophisticated analysis of protein titration data can be carried out. The electrostatic free energy for any particular ionizable group (i) can be expressed in terms of the distances, r_{ij}, between this group and all other charges on the molecule. An example of a recent application of this method to hemoglobin is given by Matthew et al. (1979). The results predict that the local charge environ-

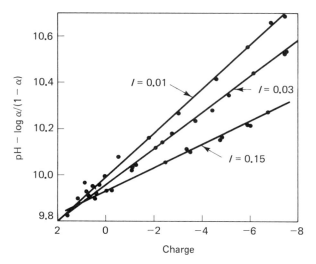

Figure 3.13 Evaluation of pK_{in} and w for the three normal tyrosines in ribonuclease. From Tanford, et al. (1955). Reprinted by permission of the American Chemical Society.

ment will have widely varying effects on local pK_a values. For example, assuming an intrinsic pK_a of 6.60 for histidine led to the prediction of values ranging from 6.07 to 8.48 for individual groups! This points up the necessity for more selective experimental methods for the study of protein titration.

Methods are being developed that allow measurement of the titration of individual classes of groups, or even individual groups in proteins and other macromolecules. For example, the absorption spectrum of tyrosine changes upon ionization; this allows us to follow the ionization of this residue independently of others. Figure 3.14 shows a titration of the tyrosyl groups in ribonuclease, as monitored by their ultraviolet absorbence. There are six tyrosines in the molecule; of these, three titrate in the range normally expected for such residues ($pK_a \simeq 10$), whereas the remaining three are abnormal, titrating near pH 12. Deprotonation of these groups involves fundamental changes in the protein structure, as shown by the irreversibility of the titration curve.

Even more detail has been provided by the use of NMR techniques, for with such it is often possible to follow the ionization of *individual* groups. For example, the four histidine groups in ribonuclease have been shown to have individual pK_a values ranging from 5.8 to 6.7. The use of such techniques demonstrates quite vividly what had been long suspected: that the fitting of protein titration curves in terms of sets of "identical" groups is a very crude approximation, which appears to succeed only because of the number of parameters at one's disposal and the limited precision of the data. In fact, it is now clear that even groups of a given kind can show a very wide range of behavior in so complicated an environment as the protein molecule.

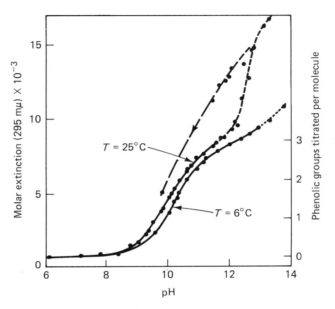

Figure 3.14 Spectrophotometric titration of the tyrosine side chains in ribonuclease. Note the two classes and evidence for irreversibility in the back titration from high pH. From Tanford, et al. (1955). Reprinted by permission of the Americal Chemical Society.

3.4 Conformation Equilibria

The equilibria between helical ordered structures and random coils appear to play important roles in biochemical processes. To cite one example, DNA appears under most circumstances as a highly ordered, double-stranded helix. Yet in the processes of replication and transcription it seems very likely that this structure must be broken down, at least temporarily or in part. Again, many native proteins contain appreciable quantities of α helix. Yet proteins are synthesized as linear chains, and must spontaneously undergo folding to form such structures. Thus it is evident that understanding of the stability of such structures is significant.

One characteristic of many such equilibria is their *all or none* character. A typical curve for the thermal transition of DNA is shown in Figure 3.15. Below 80°C no appreciable change occurs. Then, in a very narrow temperature range, the helix "melts" to a random coil. In terms of the thermodynamics, what does this mean? If we write an equilibrium constant for the reaction as

$$K = \frac{\text{fraction of residues in random-coil regions}}{\text{fraction of residues in helical regions}} \tag{3.85}$$

we can say that

$$K = e^{-\Delta G^0/RT} = e^{-(\Delta H^0 - T\,\Delta S^0)/RT} \qquad (3.86)$$

The transition can be abrupt only if ΔH^0 and ΔS^0 are both large, so that the exponent in Equation (3.86) changes from a very large negative quantity to a very large positive quantity for a small change in T. But what does this imply? We know that the ΔH and ΔS for the loosening of a single unit from the helix are modest quantities, probably of the order of 5 kcal and 20 entropy units, respectively. Such values would yield a gradual transition over the whole temperature range in Figure 3.15. A reasonable interpretation is as follows: The

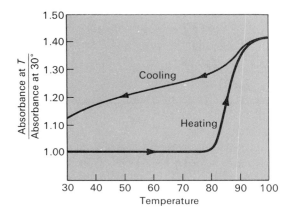

Figure 3.15 The thermal denaturation of DNA as measured spectrophotometrically.

process must be highly cooperative; that is, the entire chain (or large segments thereof) must change directly from the helix to the random-coil form. If this were the case, then the observed ΔH^0 and ΔS^0 would be the sum of all of the contributions from individual pairs. DNA should then have the kind of behavior shown in Figure 3.15. Such a change is sometimes called a *two-state transition*.

Why should these transitions be cooperative? If one examines the structure of either an α-helical polypeptide or a helical polynucleotide, it becomes evident that it is very difficult for a single residue to move out of the helix by itself. In α helices, each residue is hydrogen-bonded to a residue four units up the chain; the bonds form an interlocking set. Similarly, a base in DNA is simultaneously hydrogen-bonded to a base on the complementary chain and interacting strongly with bases in its own chain. These, in turn, are hydrogen-bonded to the complementary chain (see Figure 3.16).

The same idea can be expressed in another way: It is much more difficult to start a break in the middle of a helix than it is to release a residue at the end of a helical section. Rather sophisticated statistical-mechanical treatments of this problem have been developed. We shall not go into the details here but shall simply mention some results of one such theory, that of B. H. Zimm and J. K. Bragg (1959). In analyzing this problem, these authors distinguish between the

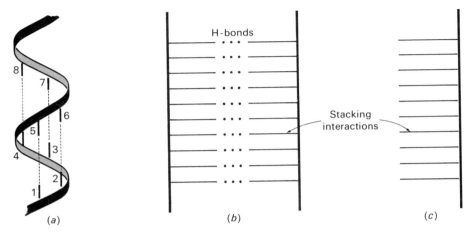

Figure 3.16 Highly schematic diagrams of (*a*) an α helix and (*b*) a double-stranded polynucleotide. Note that in both cases three-dimensional bonding structures stabilize the structure. Contrast with the single-stranded polynucleotide (*c*). The first two melt cooperatively and the third noncooperatively.

free-energy increment (ΔG_P) involved in propagating a helical region by 1 unit and that (ΔG_i) required for the initiation of a break in a helical region. The corresponding equilibrium constants are denoted by s and σs:

$$s = e^{-\Delta G_P/RT} \tag{3.87}$$

$$\sigma s = e^{-\Delta G_i/RT} \tag{3.88}$$

The detailed theory shows that if $\sigma = 0$ (that is, if ΔG_i is indefinitely large) the helix-coil transition occurs at $s = 1$ with infinite sharpness. Furthermore, as σ increases, the transition becomes broader, approaching that for a noncooperative process. This limit is approached when $\sigma = 1$. For small σ, the breadth of the transition is proportional to $\sqrt{\sigma}$. Typical results from this kind of calculation are shown in Figure 3.17. The midpoint of the transition always occurs at

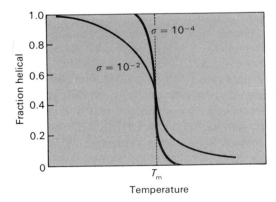

Figure 3.17 Graphs of predicted fraction helix for cooperative melting with two different values of σ.

$s = 1$ (the point where it would occur even if the transition were noncooperative). Lower values of σ simply make it sharper. A detailed analysis of the Zimm–Bragg theory is given in Cantor and Schimmel, Part III (1980).

The relationship of this to our earlier discussion is brought out by another equation:

$$\Delta H_{eff} \simeq \bar{\nu} \, \Delta H_P \qquad (3.89)$$

where ΔH_{eff} is the apparent enthalpy change measured from the van't Hoff equation (3.23), ΔH_P is the enthalpy change for release of one residue, and $\bar{\nu}$ is the average number of segments that make the transition in concert.

Lest it be thought that all conformational changes in macromolecules are of the cooperative type, consider the example in Figure 3.18. Polyriboadenylic acid,

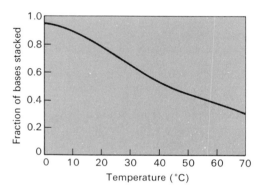

Figure 3.18 An approximate representation of the thermal denaturation of polyriboadenylic acid at neutral pH. This is a single-stranded structure.

in neutral solution at low temperatures, forms an ordered single-stranded helix. This melts gradually as shown, yielding $\Delta H^0 \simeq 9$ kcal, $\Delta S^0 \simeq 30$ e.u. These values are indistinguishable from those obtained for the dimer; thus the melting is wholly noncooperative. This should not be surprising, for the single-standed structure is stabilized only by stacking interactions of each base with its neighbors. Any such "bond" can break, without concern for the state of adjacent bonds.

The unfolding, or "denaturation" of protein molecules appears to be a highly cooperative process. One way to judge this is by comparison of calorimetric and van't Hoff enthalpies. As Privalov (1979) has pointed out in a review of protein stability, the ratio $\Delta H_{cal} / \Delta H_{eff} = 1.05 \pm 0.03$ for a number of well-studied proteins. The entries in Table 3.2 show that the enthalpy changes can be quite large compared to the energies of individual noncovalent interactions. This is to be expected for a cooperative process, which sums a large number of small contributions. What is surprising in the data of Table 3.2 is the wide range of enthalpies for unfolding processes. Some are even negative, which is not in accord with simplistic expectations. A negative enthalpy of denaturation requires that folding of the molecule involves an *entropy increase*. How can this be so?

TABLE 3.2 *Some calorimetric data on conformational transitions[a]*

Substance, Reaction	pH	Solvent	T (°C)	ΔH (kcal/mole)	Method
Trypsin, denaturation	~2	0.1 M NaCl	25	8.0	Heat of mixing
Myoglobin, denaturation	4.5	0.15 M KCl	30	40	Heat of mixing
Ribonuclease A, denaturation	2.8	0.15 M KCl	44	86.5	Heat capacity
Ferrihemoglobin, denaturation	~3.5	0.02 M Na formate	15	−76	Heat of mixing
Poly-L-glutamate, denaturation	~5	0.1 M KCl	30	−1.1/residue	Heat of mixing
Poly-α-benzyl-L-glutamate, denaturation	—	25% DCE[b] 75% DCA[b]	26	0.53/residue	Heat capacity
Salmon DNA, denaturation	6.0	0.1 M NaCl	25	8.31/base pair	Heat of mixing
Sea urchin DNA, denaturation	6.0	0.1 M NaCl	25	8.03/base pair	Heat of mixing
Poly A (double-stranded),	4.1	0.1 M KCl, 0.01 M cacodylate	25	2.7/nucleotide	Heat of mixing
Poly A (single-stranded), denaturation	7.3	0.1 M KCl, 0.01 M tris	35	9.4/nucleotide	Heat capacity
Poly (A + U) ⟶ poly A + poly U	6.6	0.1 M KCl, 0.01 M cacodylate	25	5.9/base pair	Heat of mixing

[a]Data from H. Sober (ed.), *The Handbook of Biochemistry and Molecular Biology*, Chemical Rubber Co., Cleveland, Ohio, 1968.

[b]DCE, dichloroethanol; DCA, dichloroacetic acid.

One explanation is provided by the role of hydrophobic interactions in the stabilization of native protein structure. Hydrophobic side chains, if placed in contact with water, will cause an ordering of water structure. When these residues are withdrawn from aqueous contact (as when a denatured protein folds) this ordered water is released. The entropy of the whole system (protein plus solvent) increases, and this entropy increase more that compensates for the unfavorable enthalpy change in the reaction.

The term "hydrophobic bonding" is often applied to this kind of stabilization of protein structure. However, this is in a sense a misnomer. It is not that bonds are formed between hydrophobic residues; rather, by moving to the inside of a folded structure they cause an entropy-driven stabilization of the macromolecule.

It should also be noted that an entropy increase probably makes an important contribution to the free energy of stabilization from ion-pair interactions in protein molecules. As do hydrophobic residues, charged groups on the surface of the protein will tend to order water molecules about them; the release of these water molecules will lead to an entropy increase for the system.

3.5 Macromolecular Interactions

So far, we have focused on interactions of small molecules or ions with macromolecules and on the conformational changes within biopolymers. However, many of the important biochemical processes in cells are based on, or controlled by, interactions between two or more macromolecules. Examples abound: the formation of multisubunit proteins from polypeptide chains; the intimate and complex association between protein and RNA molecules in the ribosome; the control of DNA expression by the binding of specific proteins, and so forth. Most of these interactions are based on noncovalent forces, of the kinds detailed in Table 1.1, although covalent bonds such as disulfides are sometimes involved in the stabilization of the complex.

Those macromolecular complexes stabilized by noncovalent forces are in thermodynamic equilibrium with their components, and in many cases it is possible to measure the thermodynamic parameters of these reactions. A sampling of such data is given in Table 3.3. A number of noteworthy points emerge from inspection of these data.

TABLE 3.3 *Association reactions of proteins[a]*

Protein	*Reaction Type*	ΔG^0 *(kcal/mole)*	ΔH^0 *(kcal/mole)*	ΔS^0 *(e.u./mole)*	$\Delta G^0/Unit$[b] *(kcal/mole)*
Arginosuccinase	$2P \rightleftharpoons P_2$	−10.2	46	189	−5.1
Enolase	$2P \rightleftharpoons P_2$	−9.4	−80	−221	−4.7
Glutamate dehydrogenase	$P + P_i \rightleftharpoons P_{i+1}$	−7.8	0	25	−3.9
Hemerythrin	$8P \rightleftharpoons P_8$	−46.3	0	22	−5.8
Insulin	$2P \rightleftharpoons P_2$	−5.5	−7.1	−5.5	−2.7
	$2P_2 \rightleftharpoons P_4$	−4.0	−16.3	−41	−2.0
	$P_2 + P_4 \rightleftharpoons P_6$	−3.8	49	177	−1.9
β-Lactoglobulin	$4P_2 \rightleftharpoons P_8$	−15.2	−56	−151	−3.8
Trypsin inhibitor	$2P \rightleftharpoons P_2$	−4.9	0	19	−2.4
Tryptophanase	$2P_2 \rightleftharpoons P_4$	−5.2	50	198	−2.6
Tryptophan synthetase, β chain	$2P \rightleftharpoons P_2$	−7.6	0	25	−3.8

[a]Data from Klotz et al. (1975).
[b]ΔG^0 divided by the number of molecules taking part in the reaction as written. This gives a measure of ΔG^0 per macromolecule–macromolecule bonding surface.

First, although the energies for the *individual* noncovalent interactions are quite small (Table 1.1), it is clear that they can sum, in the binding between macromolecules, to yield quite considerable values. This is a reflection of the inherently cooperative nature of such reactions; macromolecules tend to interact over binding *surfaces*, and many groups may participate simultaneously. Second, an examination of the signs and magnitudes of ΔH^0 and ΔS^0 suggests that many different patterns of noncovalent interactions can be involved. In some cases hydrophobic interactions seem to dominate; in other cases, hydrogen bonding or electrostatic forces play the major roles. Finally, although the examples in Table 3.3 only hint at this, it is an inescapable fact that most biologically significant interactions between macromolecules result in the formation of defined structure of relatively simple stoichiometry. Monomer–dimer, dimer–tetramer reactions are common; indefinite aggregation is rare.

A characteristic feature of all association–dissociation reactions is the dependence of the equilibrium on the *total* concentration of the subunits involved. This serves to distinguish such reactions from purely conformational changes, which will be concentration independent. A simple case will illustrate the principle, show how such reactions can be studied, and point out some of the problems involved. Consider, for example, the analysis of a simple monomer–dimer reaction. We have the equilibrium condition

$$2M \rightleftharpoons D$$

$$K = \frac{[D]}{[M]^2} \tag{3.90}$$

where K is the association constant, in liters per mole. If we denote by M_0 the total molar concentration of subunits in the system (irrespective of whether they be present as monomer or incorporated into dimer), we obtain the additional equation

$$M_0 = [M] + 2[D] \tag{3.91}$$

The factor 2 occurs because each mole of dimer contains 2 moles of subunits. Thus we can write

$$[D] = \frac{M_0 - [M]}{2} \tag{3.92}$$

and substitute this into Equation (3.90). The resulting quadratic equation can be solved for [M], as a function of the total concentration M_0, to give for the fraction of monomer

$$\frac{[M]}{M_0} = \frac{-1 \pm \sqrt{1 + 8KM_0}}{4KM_0} \tag{3.93}$$

Only the solution with the positive sign is physically significant. If one evaluates the limits of this expression, it is found that $[M]/[M_0] \rightarrow 1$ as $M_0 \rightarrow 0$, and $[M]/M_0 \rightarrow 0$ as $M_0 \rightarrow \infty$. Thus the reaction will tend toward total dissociation or complete dimerization as M_0 becomes very small or very large, respectively. Graphs of monomer concentrations versus total concentration are shown in Figure 3.19 for two different values of the association constant, $K = 1 \times 10^4$ liters/mole and $K = 1 \times 10^8$ liters/mole, corresponding to $\Delta G^0 = -5.5$ kcal/mole and $\Delta G^0 = -11$ kcal/mole, respectively. The graphs are shown in terms of weight concentration (mg/ml) rather than molar concentration, as-

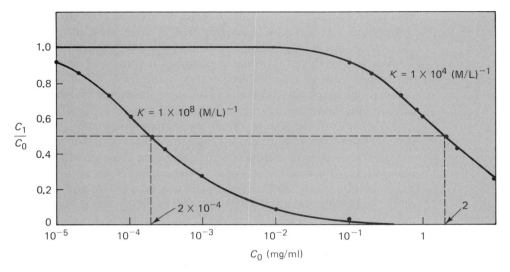

Figure 3.19 The weight fraction of monomer as a function of total protein concentration for reversible monomer–dimer association reactions. Curves for two values of the molar association constant (1×10^8 liters/mole and 1×10^4 liters/mole) are shown, with a monomer of molecular weight 20,000.

suming a molecular weight of 2×10^4 for the subunit. This is done to emphasize the following point: If $\Delta G^0 = -5.5$ kcal/mole, half of the material will be dissociated into monomer at 2 mg/ml; the reaction can be easily studied by most physicochemical techniques. But if $\Delta G^0 = -11$ kcal/mole, the binding is so strong that half dissociation only occurs at 0.0002 mg/ml. In fact, there is very little dissociation even at 0.02 mg/ml. Such strong-binding reactions are very difficult to study, since very little shift in the equilibrium occurs over the concentration range accessible to most of the methods available to the physical biochemist. The experimental techniques often used for the analysis of such association dissociation equilibria are those which yield average molecular weight values for the equilibrium mixture, such as osmotic pressure (Chapter 2), sedimentation equilibrium (Chapter 5), or light scattering (Chapter 9). A description of how such an analysis can be carried out will be given in Chapter 5.

PROBLEMS

1. (a) Using data in Table 3.1, calculate ΔG^0 for the reaction

   glucose-1-phosphate $\rightleftharpoons$ glucose-6-phosphate

   at 25°C, pH 7.0.
   (b) What is the equilibrium constant?

(c) If a mixture were prepared with 1 M glucose-6-phosphate and 1×10^{-3} M glucose-1-phosphate, which way would the spontaneous reaction go?

2. Calculate the ratio of ADP to ATP in aqueous solution (with excess Mg^{2+}) at 37°C as a function of total nucleotide concentration, from 0 to 1 M. Neglect the possibility of adenosine monophosphate formation. Assume that all phosphate comes from hydrolysis of ATP and that solutions behave ideally.

3. The data below describe the binding of a ligand A to a macromolecule. Calculate both n and the binding constant and show that all sites are identical and independent.

$[A] \times 10^3$ (*moles/liter*)	$\bar{v}$
0.5	1.6
1.0	2.5
2.0	3.2
5.0	4.0
10.0	4.1
20.0	4.8

4. The data below show the binding of oxygen to squid hemocyanin. Determine, from a Hill plot, whether the binding is cooperative and estimate the *minimum* number of subunits in the molecule.

pO_2 (*mm*)	*Percent Saturation*	pO_2 (*mm*)	*Percent Saturation*
1.13	0.30	136.7	55.7
5.55	1.33	166.8	67.3
7.72	1.92	203.2	73.4
10.72	3.51	262.2	79.4
31.71	8.37	327.0	83.4
71.87	18.96	452.8	87.5
100.5	32.9	566.9	89.2
123.3	47.8	736.7	91.3

5. Consider a protein that contains a *single* binding site which can exist in two forms, R and T, with binding constants k_R and k_T. These are assumed to be in equilibrium, with $L = [T]/[R]$.

(a) Prove that this protein will exhibit noncooperative binding, despite the T $\rightleftharpoons$ R equilibrium.

(b) What is the apparent binding constant, k, in terms of L, k_R, and k_T?

*6. (a) Show that the fraction of protein in the R state is given, by the MWC model, as

$$\bar{R} = \frac{(1 + \alpha)^n}{(1 + \alpha)^n + L(1 + c\alpha)^n}$$

where $\alpha = k_R[A]$.

(b)   Since α can only vary from 0 to ∞, there are limits to the range of values possible for $\bar{R}$. This is called the "allosteric range." Find these limits, in terms of L and c.

(c)   What is the allosteric range for a tetramer ($n = 4$) when $c = 10^{-2}$, $L = 10^4$ and when $c = 10^{-2}$, $L = 10^8$?

7.   In a simplified version of the MWC model, it is assumed that ligand binding can occur *only* to the R state. Consider a molecule with two binding sites which obeys this model.

(a)   Write down the equation for $\bar{v}$ for this model.

(b)   Compute a Hill plot for $L = 6$, $k_R = 10^6$ for this model. What is the maximum slope?

*(c)   Derive an expression in terms of L for the maximum Hill slope.

8.   (a)   The molecules of a homogeneous sample of a copolymer each contain 100 aspartic acid residues ($pK \simeq 4.5$) and 50 histidine residues ($pK \simeq 7$). Neglecting electrostatic interaction, calculate the titration curve.

(b)   Sketch an accompanying curve to show qualitatively the effects one would expect from electrostatic interaction between the sites.

*9.   Consider the transition from an α helix to a random coil for a polypeptide containing n amino acid residues. Assume that the transition is cooperative.

(a)   Assuming that one bond in each residue in the random coil may take one of *three* equiprobable orientations, obtain an equation for the number of configurations and hence the entropy change in going from 1 mole of helices (each of which has one configuration) to 1 mole of coils, each with many. [*Note:* To specify a configuration, you may regard the first two bonds as fixed.]

(b)   If the structure (α helix) is to be stable, we must have $\Delta G = 0$ somewhere above room temperature, (say at 50°C). If the enthalpy change for breaking 1 mole of hydrogen bonds is called ΔH_{res}, obtain an expression for ΔG in terms of n and calculate what ΔH_{res} must be for $\Delta G = 0$ at 50°C, if $n = 100$. [*Note:* Be careful about the situation at the helix ends in relating ΔH to ΔH_{res}.]

(c)   Using the "exact" expression for ΔG, show that the melting temperature, defined as the T for which $\Delta G = 0$, decreases with decreasing n (that is, short helices are less stable than long ones).

10.   Take the data in Table 3.3 and plot ΔS^0 versus ΔH^0. What do you find? Discuss possible implications of this observation.

11.   The subunits of the hemocyanin of the shrimp, *Callianassa*, can undergo the following association reactions:

$$2P \;\rightleftharpoons\; P_2 \qquad K_{12} = \frac{[P_2]}{[P]^2} \tag{1}$$

$$2P_2 \;\rightleftharpoons\; P_4 \qquad K_{24} = \frac{[P_4]}{[P_2]^2}$$

$$\text{or} \quad K_{14} = \frac{[P_4]}{[P]^4} \tag{2}$$

Equilibrium constants were measured as a function of T:

$T\,(^\circ C)$	K_{12}	K_{14}
4	1.3×10^7	4.5×10^{20}
10	1.2×10^7	1.0×10^{21}
15	1.5×10^7	3.0×10^{21}
30	—	4.0×10^{22}

All concentrations were expressed in moles per liter. Estimate ΔH^0, ΔS^0, and ΔG^0 at 20°C for reactions (1) and (2).

REFERENCES

General

Denbigh, K. G.: *The Principles of Chemical Equilibrium*, Cambridge University Press, Cambridge, 1955.

Eisenberg, D. and D. Crothers: *Physical Chemistry with Applications to the Life Sciences*, The Benjamin-Cummings Publishing Co., Menlo Park, Calif., 1979, Chaps. 4 and 5.

Binding

Koshland, D. E., G. Nemethy, and D. Filmer: *Biochemistry*, 5, 365–385 (1966). The primary reference to the KNF theory.

Monod, J., J. Wyman, and J. P. Changeux: *J. Mol. Biol.*, **12**, 88–118 (1965). The original paper on the MWC theory of allostery.

Steinhardt, J. and S. Beychok: "Interaction of Proteins with Hydrogen Ions and Other Small Ions and Molecules," in *The Proteins*, 2nd ed. (H. Neurath, ed.), Academic Press, Inc., New York, 1964, Vol. II, pp. 139–304. Still a useful review on protein titration.

Weber, G.: "Energetics of Ligand Binding to Proteins," *Adv. Protein Chem.*, **29**, 1–83 (1975). A general treatment that does not resort to particular models.

Detailed Studies of Titration of Proteins

Matthew, J. B., G. I. H. Hanania, and F. R. N. Gurd: *Biochemistry*, **18**, 1919–1928 (1979).

Roberts, G. C. K. and O. Jardetsky: "Nuclear Magnetic Resonance Spectroscopy of Amino Acids, Peptides, and Proteins," *Adv. Protein Chem,*. **24**, 447–545 (1970).

Tanford, C. and J. D. Hauenstein: *J. Am. Chem. Soc.*, **78**, 5287–5291 (1956).

Tanford, C., J. D. Hauenstein, D. G. Rands: *J. Am. Chem. Soc.*, **77**, 6409–6413 (1955).

Conformational Equilibria

Cantor, C. R. and P. R. Schimmel: *Biophysical Chemistry*, W. F. Freeman and Company, Publishers, San Francisco, 1980, Chap. 20.

Privalov, P. L.: "Stability of Proteins," *Adv. Protein Chem.*, **33**, 167–241 (1979).

Zimm, B. H. and J. K. Bragg: *J. Chem. Phys.*, **31**, 526–535 (1959). A pioneering paper in this field.

Macromolecular Interactions

Klotz, I. M., D. W. Darnall, and N. R. Langerman: "Quaternary Structure of Proteins," in *The Proteins*, 3rd ed. (H. Neurath and R. L. Hill, eds.), Academic Press, Inc., New York, 1975, Vol. I, pp. 293–411. An excellent review of structures and the energetics of subunit interactions.

Chapter Four

INTRODUCTION TO

TRANSPORT PROCESSES:

DIFFUSION

So far we have considered almost exclusively the study of systems that are at equilibrium. However, for all their rigor and utility, such studies are quite uninformative about details of molecular structure. The reason for this should be clear. Thermodynamics is a science of the macroscopic properties of matter, and can tell us little about how individual molecules behave, or are conformed. The biochemist who needs such information will often turn to techniques such as sedimentation, diffusion, or electrophoresis. These are *irreversible processes* that occur only when a system has been removed from a state of equilibrium; they represent the relaxation of the system toward a new equilibrium. We shall see that while thermodynamics can be extended to a certain extent to describe these phenomena, their detailed analysis necessarily involves nonthermodynamic concepts concerning the behavior of individual molecules.

There is a second aspect to transport processes which is of great importance to the biochemist. Some of these techniques, such as sedimentation and electrophoresis, can be used for the analytical or preparative *separation* of macromolecules. We shall also consider these applications in subsequent chapters.

4.1 General Features of Transport Processes

There are two very different ways in which we can attempt to analyze transport processes. We may try to see how molecules will behave, as particles subjected to external forces exerted upon them. Such a "mechanical" treatment will suffer from considerable imprecision, for it is obviously impossible to describe the moment-to-moment experiences of a solute molecule. Nevertheless, even a rough treatment will give a picture of what happens. Alternatively, the methods of equilibrium thermodynamics may be extended to cover some systems that are not too far from equilibrium. We may expect that in this way some of the rigor that accompanies thermodynamics may carry over, but since thermodynamics does not depend on the existence of molecules, the details of the picture will be lost. Both of these points of view will be used in succeeding chapters.

Taking at first the mechanical point of view, let us imagine in a solution a solute molecule to which an external force F_i is suddenly applied. An electric field might be switched on or a centrifugal force imposed by spinning the solution. Whatever its random motion may be, the molecule will also now be accelerated in the direction of the field. But this acceleration will last for only an *exceedingly* short time† (see Figure 4.1 and Problem 1), for as its velocity

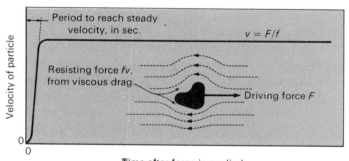

Figure 4.1 A schematic picture of the behavior of a particle suddenly subjected to a force *F*.

increases, the molecule will experience an increasing frictional resistance to motion through the medium. This frictional force will be given by $f_i v_i$, where v_i is the velocity and the constant f_i is called the *frictional coefficient* of the molecule.

A constant velocity will be reached when the total force on the molecule

---

†The student can be convinced that this time will be of the order of nanoseconds (10^{-9} sec) by solving the equation of motion for a spherical particle falling through a viscous medium (see Problem 1).

is zero:

$$f_i v_i - F_i = 0 \qquad (4.1)$$

Thus, if we can measure the velocity of motion produced by a known force, we can determine the frictional coefficient. This is *one* of the reasons for the study of transport processes, for f_i depends on the size and shape of the molecule.

The frictional resistance arises because solvent must flow around the moving object, and energy must be expended due to the viscosity of the fluid. The calculation of frictional coefficients is a difficult exercise in hydrodynamic theory, and exact expressions have been obtained for only a few simple particle shapes. For example, for a sphere of radius R

$$f_0 = 6\pi\eta R \qquad (4.2)$$

where η is the viscosity of the medium. Equation (4.2) is called *Stokes' law*. It is found that for a given particle volume, a sphere will have the minimum possible frictional coefficient, which we call f_0. Therefore, for any of the other particle models listed in Table 4.1, we list the ratio (f/f_0) of the frictional coefficient of the particle to that of a sphere of equal volume. This *frictional ratio* is always $\geqslant 1$, as shown in Figure 4.2.

TABLE 4.1   *Frictional coefficient ratios[a]*

Shape	f/f_0	R_e
Prolate ellipsoid	$\dfrac{P^{-1/3}(P^2 - 1)^{1/2}}{\ln[P + (P^2 - 1)^{1/2}]}$	$(ab^2)^{1/3}$
Oblate ellipsoid	$\dfrac{(P^2 - 1)^{1/2}}{P^{2/3}\tan^{-1}[(P^2 - 1)^{1/2}]}$	$(a^2b)^{1/3}$
Long rod	$\dfrac{(2/3)^{1/3}P^{2/3}}{\ln 2P - 0.30}$	$\left(\dfrac{3b^2a}{2}\right)^{1/3}$

[a]In these equations, $P = a/b$, where a is the semimajor axis, or the half-length for a rod (see Figure 2.1). R_e is the radius of a sphere equal in volume to the ellipsoid or rod, so $f_0 = 6\pi\eta R_e$.

Although we shall find this kind of analysis of transport processes to be very useful in subsequent chapters, it must be realized that it is at best an approximate treatment. We cannot always precisely define the force applied to an individual molecule by a known external field, and equations such as Stokes' law have been derived only for the motion of particles through continuous media. The actual bumping and jostling of molecules is neglected entirely. However, the fact that equations such as (4.2) depend on molecular dimensions is one reason for the importance of transport processes in the study of macro-molecules (see Section 5.1).

A much more elegant framework for the description of transport processes

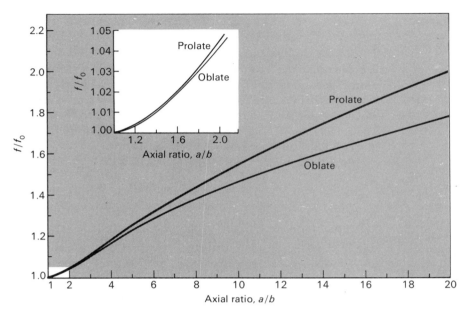

Figure 4.2   The dependence of frictional coefficient on particle shape. The ratio f/f_0 is the frictional coefficient of an ellipsoid of the given axial ratio divided by the frictional coefficient of a sphere of the same volume as the ellipsoid.

has been developed in recent years. This is the *thermodynamics of irreversible processes*, which allows the discussion of systems not too far from equilibrium in much the manner of classical thermodynamics. We shall not attempt to develop this theory here but shall use some of its methods to give a more exact and coherent picture of transport processes. There is an altogether different reason the biologist and biochemist should be acquainted with this theory. The life of cells and organisms is an irreversible process, at all levels.

   The phenomena that we shall discuss (diffusion, sedimentation, electro-phoresis, and so forth) have one feature in common: A system *not* in equilibrium moves towards equilibrium. The final state is dictated, of course, by the temperature, pressure, composition, and external forces imposed on the system. During the drive toward equilibrium, *flow* must occur. In the problems of interest to us, this is a flow of matter—as, for example, in the movement of molecules in a centrifugal field. We define the flow (J_i) of a component i as the number of mass units (grams, moles, and so forth) of i crossing 1 cm² of surface in 1 sec. The flow is obviously related to the velocity of transport. If solute molecules in a region of concentration C_i are all moving with a velocity v_i in a direction perpendicular to a surface s (see Figure 4.3), the flow and velocity are

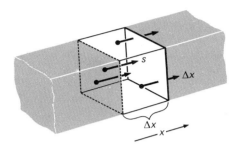

Figure 4.3 The relation between molecular velocity and flow. In time Δt a molecule with velocity v moves a distance $\Delta x = v\Delta t$. Thus, all the solute molecules in a slab going back this distance from the surface s will pass through s in Δt seconds. The amount passing will be the concentration times the slab volume $\Delta w = Cs\,\Delta x$, or $\Delta w = Csv\,\Delta t$. Since the flow is defined as $J = \Delta w/s\,\Delta t$, we obtain $J = Cv$.

related by

$$J_i = v_i C_i \tag{4.3}$$

What, precisely, drives the flow when a system is moving toward equilibrium? From the point of view of classical mechanics, we would say that forces are acting on the molecules. However, in the theory of irreversible processes, we do not attempt to describe the forces on individual molecules but take a more general view. In classical mechanics, a force is always the negative of the gradient (rate of change with distance) of a potential energy: similarly we shall identify the generalized "forces" that drive the processes of diffusion, sedimentation, and electrophoresis with the gradients of certain "potentials." When these potentials become uniform throughout the system, no force exists, and flow no longer occurs.

An example is familiar to the reader: If potential differences exist in an electrical circuit, current flows until these differences are eliminated. The similarity of this and the processes we are discussing is pointed up in Table 4.2.

TABLE 4.2   *Transport processes*

Process	*Potential*	*Flow of:*	*State of Equilibrium*
Electrical conduction	Electrostatic	Electrons	Uniform electrostatic potential
Heat conduction	Temperature	Heat	Uniform temperature
Diffusion	Chemical potential	Molecules	Uniform chemical potential
Sedimentation	Total potential = chemical potential + centrifugal potential energy	Molecules	Uniform total potential

A primary result of the theory of irreversible processes is the statement that the flow, at any point in the system, at any instant, is proportional to the gradient in the appropriate potential, there and then. For one-dimensional flow in an ideal system,

$$J_i = -L_i \frac{\partial U_i}{\partial x} \tag{4.4}$$

This equation can be thought of as a generalization of Ohm's law with U_i a generalized potential and L_i a generalized "conductivity."† It says that the farther the system is from equilibrium, the faster it moves toward it. Equation (4.4) is written for potentials that change in one direction only; the extension to three dimensions is easy, but then both the gradient and the flow *must* be written as vector quantities.

We can now connect the mechanical and thermodynamic analyses of transport phenomena. We write Equation (4.1) for a particular component i, and consider the force per molecule F_i as $-(1/\mathfrak{N})(\partial U_i/\partial x)$, where $\mathfrak{N}$ is Avogadro's number. Using Equation (4.3), then (4.4) may be rewritten as

$$C_i v_i = \mathfrak{N} L_i F_i \qquad (4.5)$$

or

$$v_i = \frac{\mathfrak{N} L_i}{C_i} F_i \qquad (4.6)$$

which says that $L_i = C_i/\mathfrak{N} f_i$, $\mathfrak{N} f_i$ being the frictional coefficient per mole.‡ It is well to remember that the two theories start from different points of view and that the f_i defined from the thermodynamic theory may not be *exactly* the same quantity that we would obtain from a naive calculation from molecular dimensions.

Before turning to detailed description of particular transport processes, there is one more general point to consider. The experimenter rarely measures flow directly; more often, the changes in concentration with time in various parts of the system are followed. Thus we must relate flow to these concentration changes. This can be accomplished by utilizing the fact that matter is not being created or destroyed in the system, in other words, the conservation of mass. A cell of uniform cross section (A) in which a flow of solute is occurring is shown in Figure 4.4. Consider a thin slab, perpendicular to the direction of flow, with surfaces at x and $x + \Delta x$. How will the concentration change in this volume element in time Δt? In the general case, we shall assume that J_i may be different at different points. Then the net change in the mass of solute in the slab (Δw_i) must be given by the flow *in* $[J_i(x)]$ minus the flow *out* $[J_i(x + \Delta x)]$, each multiplied by the area and Δt:

$$\Delta w_i = J_i(x)A\,\Delta t - J_i(x + \Delta x)A\,\Delta t \qquad (4.7)$$

†Equation (4.4) is for a relatively simple case, where one kind of potential gradient and one kind of flow exist. If one has gradients in several potentials and flow of several quantities (several chemical components, heat, and so forth), the general result is $J_i = -\sum_k L_{ik}(\partial U_k/\partial x)$. This says that the flows may be *coupled*; the existence of a gradient in potential k may cause a flow of substance i.

‡The relationship $L_i = C_i/\mathfrak{N} f_i$ may be thought of in the following way: It says that the *mass conductivity* (L_i) is proportional to the concentration of the substance being transported and inversely proportional to the resistance the medium offers to that transport.

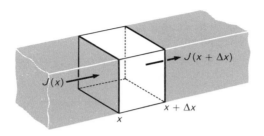

Figure 4.4 The continuity equation. The change of concentration in the slab between x and $x + \Delta x$ depends upon the difference between the flow in and the flow out.

If we divide this by the volume of the slab, we obtain the change in concentration:

$$\Delta C_i = \frac{\Delta w_i}{A \, \Delta x} = \frac{J_i(x) - J_i(x + \Delta x)}{\Delta x} \, \Delta t \tag{4.8}$$

or, rearranging,

$$\frac{\Delta C_i}{\Delta t} = \frac{J_i(x) - J_i(x + \Delta x)}{\Delta x} \tag{4.9}$$

When the increments become infinitesimal, we get the differential equation of continuity

$$\frac{\partial C_i}{\partial t} = -\frac{\partial J_i}{\partial x} \tag{4.10}$$

In words, this simply says that the concentration will increase in a region if material is coming in faster than it is going out. Obviously, the equation will have to be modified if chemical reactions are making or using up substance i or if A depends on x.

4.2 Diffusion

Let us now apply the generalities that have been developed to a specific transport process—diffusion. The arrangement shown in Figure 4.5 is one way in which diffusion can be studied. By one or another mechanical trick, solvent is layered

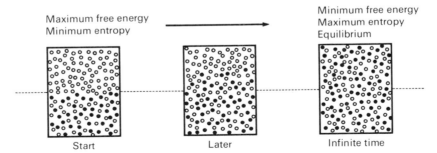

Figure 4.5 The process of diffusion. Open circles are solvent molecules; filled circles are solute.

carefully on a solution. This gives a sharp *boundary*, across which the solute concentration changes abruptly from zero to a value C_0. Common sense indicates that this situation cannot persist, for the random Brownian motion of the molecules will blur the boundary and eventually lead to complete mixing. Thermodynamics says the same thing in a different way; the initial state is certainly not one of equilibrium, for the free energy will be lower when uniform mixing has occurred. Neglecting small contributions from the heat of mixing, the free-energy change results entirely from the higher entropy of the mixed state.

In fact, our previous discussion of the thermodynamics of solutions gives a very precise prescription for the equilibrium state: The chemical potential must be constant throughout the system. Thus the change of chemical potential with distance in the boundary region is the "driving force" for the flow of solute that will lead to equilibrium:

$$J_2 = -L_2 \frac{\partial \mu_2}{\partial x} \tag{4.11}$$

The subscript 2 denotes the solute component. The chemical potential of the solute depends on temperature, pressure, and the concentration, but since T and P are assumed to be uniform,

$$\frac{\partial \mu_2}{\partial x} = \left(\frac{\partial \mu_2}{\partial C_2}\right)_{T,P} \frac{\partial C_2}{\partial x} \tag{4.12}$$

Thus under these conditions the concentration gradient determines the flow. From Equation (2.33), $(\partial \mu_2/\partial C_2)_{T,P} = (RT/C_2)[1 + C_2(\partial \ln y_2/\partial C_2)]$, where y_2 is the activity coefficient. Then

$$J_2 = -\frac{L_2 RT}{C_2}\left(1 + C_2 \frac{\partial \ln y_2}{\partial C_2}\right) \frac{\partial C_2}{\partial x} \tag{4.13}$$

The factor L_2/C_2 in Equation (4.13) can be replaced by $1/\mathfrak{N}f_2$, bringing in the frictional coefficient:

$$J_2 = -\frac{RT}{\mathfrak{N}f_2}\left(1 + C_2 \frac{\partial \ln y_2}{\partial C_2}\right) \frac{\partial C_2}{\partial x} \tag{4.14}$$

or

$$J_2 = -D\frac{\partial C_2}{\partial x} \tag{4.15}$$

where

$$D = \frac{RT}{\mathfrak{N}f_2}\left(1 + C_2 \frac{\partial \ln y_2}{\partial C_2}\right) \tag{4.16}$$

The quantity D is termed the *diffusion coefficient*. From an experimental point of view it is defined by Equation (4.15), called *Fick's first law*. This equation is in accord with our intuitive expectation that the flow will cease only when the concentration is uniform. Equation (4.16) asserts that D depends on RT (which may be taken as a measure of the kinetic energy of the molecules); the correction term $[1 + C_2(\partial \ln y_2/\partial C_2)]$, which expresses the fact that the chemical potential depends on solute–solute interaction, and the size and shape of the molecule, which will influence the frictional coefficient. It is the last quantity that makes the macromolecular chemist interested in diffusion. For ideal solutions we may neglect the activity coefficient factor and obtain $D = RT/\mathfrak{N}f_2$. If, in addition, f_2 does not vary with concentration, D becomes a constant.

Equation (4.15) does not suggest an easy method for the determination of D. We would rather know how the concentration changes with time at various points in the boundary. This can be obtained by combining Equation (4.15) with the continuity equation (4.10):

$$\frac{\partial C_2}{\partial t} = -\frac{\partial J_2}{\partial x} \quad \text{and} \quad J_2 = -D\frac{\partial C_2}{\partial x}$$

$$\frac{\partial C_2}{\partial t} = D\frac{\partial^2 C_2}{\partial x^2} \tag{4.17}$$

Here the assumption has been made that D does not change with concentration, usually a satisfactory approximation.

Equation (4.17), which is called *Fick's second law*, is very much like the partial differential equation for heat conduction. It may be solved for specific cases if we specify initial and boundary conditions. Figure 4.6 shows a common experimental arrangement for diffusion studies. The solution column is quite long, so that it may be assumed that the concentrations at the top and bottom ($x = +\infty$ and $-\infty$) remain zero and C_0, respectively, as long as the process is followed. A sharp boundary is formed at $x = 0$ at time $t = 0$ (see Chapter 6 for

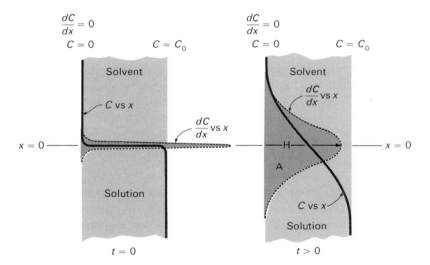

Figure 4.6 The spreading of an initially sharp boundary during free diffusion. The heavy line traces the concentration as a function of height in the cell [Equation (4.18)]. The dashed line shows the concentration gradient, according to Equation (4.19).

the forming of such boundaries). Under these conditions of *free diffusion*, the solution of Equation (4.17) is

$$C_2 = \frac{C_0}{2}\left(1 - \frac{2}{\sqrt{\pi}} \int_0^{x/2(Dt)^{1/2}} e^{-y^2}\, dy\right) \tag{4.18}$$

where y is simply a dummy variable of integration. This equation gives the kind of S-shaped curve shown in Figure 4.6 (see Problem 5).

Equation (4.18) is awkward to use or visualize, but its derivative (the concentration gradient) is simply a Gaussian "error" curve:

$$\frac{\partial C_2}{\partial x} = \frac{C_0}{2(\pi Dt)^{1/2}} e^{-x^2/4Dt} \tag{4.19}$$

It turns out that the gradient is easily measured by use of a Schlieren optical system (see Section 5.2), and D can be obtained in this way. However, the method is rarely employed at the present time, for more rapid or accurate techniques have been developed.

A more accurate method for the measurement of D utilizes a Rayleigh interferometer to determine the concentrations at different points in the cell (see Section 5.2). This requires, of course, the use of Equation (4.18). Essentially, the calculation proceeds as follows: A particular concentration, at a specific time, corresponds to a specific value of $x/2(Dt)^{1/2}$. The value of x at which this concentration exists can be measured from the position of a given fringe in the interference pattern. Thus D is determined at a whole series of points at the time of each photograph. By this procedure, average values of D with a precision of about ± 0.1 percent can be determined. An even more accurate interferometric technique called the *Gouy method* uses the symmetrical boundary itself to produce an interference pattern. So far, this method has been used mainly for low-molecular-weight materials, since interpretation of the patterns is easy only for rigorously Gaussian boundaries. Some results obtained by these methods are summarized in Table 4.3.

Equation (4.18) represents a solution of Fick's second law [Equation (4.17)] for the initial condition of a sharp *boundary* between solvent and solution. As such, it is of importance in understanding such processes as moving-boundary sedimentation (see Chapter 5). But if we wish to see the effects of diffusion on narrow *zones*, as in zonal sedimentation (Chapter 5) or zonal electrophoresis (Chapter 6), a different solution of Equation (4.17) is required. In such experiments the initial condition is a very narrow zone of concentrated solution, surrounded on both sides by buffer. As depicted in Figure 4.7, we denote the width of the initial zone by δ and the concentration therein by C_0. The appropriate solution of Equation (4.17) is now

$$C_2 = \frac{C_0\delta}{2\sqrt{\pi Dt}} e^{-x^2/4Dt} \tag{4.20}$$

TABLE 4.3 *Diffusion coefficients[a]*

Substance	Molecular Weight	$D_{20,w} \times 10^6$ [b]	Method[c]
Glycine[d]	75	9.335	G
Sucrose[d]	342	4.586	G
Ribonuclease	13,683	1.068	R
Serum albumin[d] (bovine)	66,500	0.603	R
Tropomyosin	93,000	0.224	S
Fibrinogen[d] (human)	330,000	0.197	R
Myosin[d]	440,000	0.105	S
Tobacco mosaic virus	About 40,000,000	0.053	S
Robbit papilloma virus	About 47,000,000	0.059	S

[a]The diffusion coefficients have been corrected to water at 20°C (see Chapter 5 for the procedure).

[b]Note that D generally decreases with molecular weight but that elongated molecules such as tropomyosin, fibrinogen, myosin, and TMV have unusually low values. The dimensions of D are cm²/sec.

[c]G, Gouy; R, Rayleigh; S, Schlieren.

[d]Extrapolated to zero concentration.

The reader may verify that this is a solution by showing that Equation (4.20) satisfies (4.17). The solution given is only approximate at small x (comparable to δ), but it clearly satisfies the boundary conditions, $C_2 = 0$ as $x \rightarrow \pm\infty$. As one might expect, the zone will be of Gaussian shape, and spread with time.

4.3 Diffusion and Brownian Motion

So far, we have described diffusion from a macroscopic viewpoint. But intuition tells us that the microscopic mechanism must involve nothing more than the random wandering of solute molecules caused by the random impact of solvent molecules in thermal motion. How can these pictures be reconciled?

Imagine a thin zone of solution, as in Figure 4.7, and consider the path of a single typical solute molecule as it wanders out of this zone (Figure 4.8). The path can be described as a series of short steps, each randomly oriented with respect to the preceding step. We ask: What is the mean-square displacement in the x direction (since we are considering diffusion only in the x direction) after n steps? For a given path, we can draw a vector from the origin to the final point; this vector $\mathbf{h}$ will be the vector sum of vectors corresponding to the individual steps ($\mathbf{l}_i$)

$$\mathbf{h} = \sum_1^n \mathbf{l}_i \tag{4.21}$$

To obtain the magnitude of the square of the vector, we simply multiply it

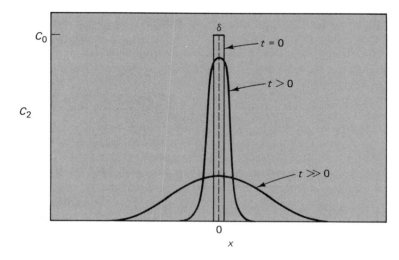

Figure 4.7 Diffusion from a narrow zone.

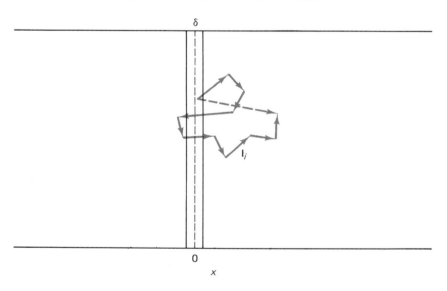

Figure 4.8 Diffusion from a narrow zone as viewed in terms of the Brownian motion of one molecule.

by itself; that is, we take the "dot" product of the vector with itself:

$$h^2 = \mathbf{h} \cdot \mathbf{h} = \left(\sum_1^n \mathbf{l}_i\right) \cdot \left(\sum_1^n \mathbf{l}_j\right)$$

$$= \sum_i \sum_j \mathbf{l}_i \cdot \mathbf{l}_j \tag{4.22}$$

The summation in Equation (4.22) represents all possible products between individual step vectors. These products will be of two kinds: When $i = j$, we are multiplying a vector by itself, and will simply get l_i^2. But if we multiply two different steps, we obtain $l_i l_j \cos\theta$, where θ is the angle shown in Figure 4.8. If we consider a large number of paths, these terms will be equally likely to be positive or negative, so they will drop out. We find, then, for the *average*-square end-to-end distance for a large number of paths,

$$\overline{h_0^2} = \sum_i^n l_i^2 = n\overline{l^2} \tag{4.23}$$

Thus the mean-*square* end-to-end distance for a random walk is proportional to the number of steps. This is the reason for the statement made in Chapter 2 that linear dimensions of a random-coil polymer are proportional to the square root of the number of segments. Such a polymer can be regarded as a random walk.

We are concerned here only with the x component of the mean-square displacement. Since the sum of the squares of the x, y, and z components must equal $\overline{l^2}$, and since all must be equal, $\overline{l_x^2} = \overline{l^2}/3$. Then

$$\overline{x^2} = n\overline{l_x^2} = \frac{n\overline{l^2}}{3} \tag{4.24}$$

where $\overline{l_x^2}$ is the mean-square x component of a random step. We can now connect this result directly to diffusion if we note that n is the product of the average step frequency (ν) and the time:

$$\overline{x^2} = \frac{\nu\overline{l^2}t}{3} \tag{4.25}$$

Equation (4.25) states that the average *square* of the displacement goes as the first power of time; the *root-mean-square displacement* $(\overline{x^2})^{1/2}$ will increase as the square root of time. This is characteristic of diffusive motion, as opposed to steady rectilinear motion, where displacement will increase as the first power of time.

We can now easily relate Equation (4.25) to results obtained from the macroscopic treatments. Equation (4.20) also described diffusion from a thin layer. We can calculate the average-square displacement of matter by

$$\overline{x^2} = \frac{\displaystyle\int_0^\infty x^2 C_2(x, t)\, dx}{\displaystyle\int_0^\infty C_2(x, t)\, dx} \tag{4.26}$$

Inserting Equation (4.20), we have

$$\overline{x^2} = \frac{\int_0^\infty \dfrac{x^2 C_0 \delta}{2\sqrt{\pi D t}} e^{-x^2/4Dt}\, dx}{\int_0^\infty \dfrac{C_0 \delta}{2\sqrt{\pi D t}} e^{-x^2/4Dt}\, dx}$$

$$= \frac{\int_0^\infty x^2 e^{-x^2/4Dt}\, dx}{\int_0^\infty e^{-x^2/4Dt}\, dx} \tag{4.27}$$

The integrals in Equation (4.27) are found in any good table of integrals; the result is

$$\overline{x^2} = 2Dt \tag{4.28}$$

Equations (4.28) and (4.25) are of exactly the same form; they become identical if

$$D = \frac{\nu \overline{l^2}}{6} \tag{4.29}$$

Note that dimensions are also correct; D (cm²/sec) is the product of a frequency (sec^{-1}) and a distance squared (cm²).

We have thus seen that the macroscopic process of diffusion can be identified with the microscopic Brownian motion—the irregular movements of individual molecules. This correspondence forms the basis for a new and rapid method for the determination of D, the *dynamic light-scattering technique*. We shall discuss light scattering in detail in Chapter 9, but for now the illustration in Figure 4.9 will suffice to explain the basis of the method. Consider a small element of volume in a solution of a macromolecule, illuminated by an intense, monochromatic light, as from a laser. Clearly, the light scattered from this volume at any instant will depend on the concentration of macromolecules contained therein. But if the volume element is small, this concentration is not constant; molecules wander in and out. Thus, at successive small intervals of time, the scattering will *fluctuate*, as depicted in Figure 4.9. The rapidity of this fluctuation will depend on how fast the molecules move. Large, slowly diffusing particles will produce slow fluctuations; small, rapidly moving molecules will cause rapid fluctuations in the scattering intensity. The fluctuations can be recorded, and analyzed in a number of ways. One of the simplest methods is to use the *autocorrelation function*. Suppose that we define the difference in the intensity at time t from the average intensity as

$$\Delta i(t) = i(t) - \bar{i} \tag{4.30}$$

Now from the intensity at some later time, $t + \tau$, we obtain $\Delta i(t + \tau)$. We construct the average of the product of these values over many such products.

$$A(\tau) = \overline{\Delta i(t)\, \Delta i(t + \tau)} \tag{4.31}$$

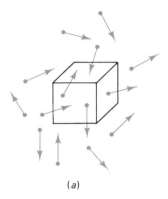

(a)

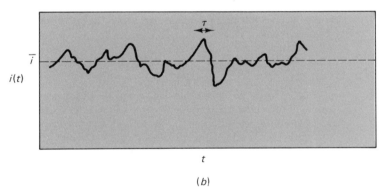

(b)

Figure 4.9 Fluctuations in light scattering can be used to measure diffusion. (a) A small element of volume from which scattering is being recorded. (b) The intensity fluctuates because solute molecules are continually wandering into and out of the element.

If τ is very large, $A(\tau)$ will approach zero, for there will be no correlation between $\Delta i(t)$ and $\Delta i(t + \tau)$. Positive and negative contributions to the average will be equally probable. On the other hand, if τ is very small, the correlation will be high, and the average will have a positive value. This can be seen by noting that for a τ much smaller than the fluctuation time, both $\Delta i(t)$ and $\Delta i(t + \tau)$ will tend to have the same sign; their product will be positive. The rate a which $A(\tau)$ decays with increasing τ will depend on how rapid the fluctuations actually are; if they are very slow, even large τ values will yield positive correlations. Thus it seems reasonable that the time decay of $A(\tau)$ should be related to the diffusion coefficient; in fact, the relationship is

$$A(\tau) = A(0)e^{-\tau/\tau_0} \tag{4.32}$$

$$\tau_0 = \frac{1}{2K^2 D} \tag{4.33}$$

$$K = \frac{2\pi n}{\lambda_0} \sin \frac{\theta}{2} \tag{4.34}$$

where n is refractive index, λ the wavelength of the light, and θ the scattering angle (see Chapter 9). The diffusion coefficient can be readily obtained from a graph of $\ln A(\tau)$ versus τ.

The dynamic light-scattering technique has a number of advantages over the older "free diffusion" methods described in Section 4.2. It is much faster (minutes rather than many hours) and requires much less sample. Although it is not yet as accurate as the interferometric measurements from free diffusion, the convenience is such that it has largely supplanted such methods.

PROBLEMS

1. Consider a small spherical particle of mass m, initially at rest, which is acted on by a constant force F turned on at $t = 0$. Solve the equation of motion for this particle. which must be $F - fv = m\, dv/dt$, where v is the velocity. Show that it approaches a constant velocity $v_{\max} = F/f$. If the particle is spherical, of radius 100 Å, and of density 1.5 g/cm^3, calculate the time required to attain 99 percent of the final velocity.

*2. Two large compartments, A and B, are separated by a porous disk or membrane of thickness Δx, through which a solute can diffuse. The solute is initially present at different concentrations, C_A and C_B in the two compartments. These are stirred, and diffusion is allowed to proceed through the disk. If A and B are large, a steady state will be obtained *in the disk*. Show that the gradient in the disk will be linear:

$$C(x) = C_A + \frac{(C_B - C_A)x}{\Delta x}$$

3. Using the result of Problem 2 and Fick's first law, derive an expression for the change of $C_B - C_A$ with time, assuming that A and B each have volume v and that the porous disk has an effective area s. This is the principle of the porous-plate method for measurement of D.

4. Obtain an equation for the half-width of a free diffusion boundary—the value of x for which $(\partial C/\partial x) = H/2$. Calculate the half-width of boundaries of sucrose and TMV after 1 hr of diffusion at 20°C.

*5. Obtain Equation (4.19) and then (4.18) by solution of Fick's second law [Equation (4.17)] subject to the boundary conditions $C = 0$ ($x = +\infty$), $C = C_0$ ($x = -\infty$). [*Hint:* This is difficult unless you make the guess that the solution will be of the form $C = C(u)$, where $u = x/t^{1/2}$. Choice of this new variable reduces the partial differential to a simple differential equation in u. Another substitution, $y = dc/du$, allows easy integration.]

6. Frictional coefficient data are frequently informative about molecular shape. From the data given in Table 4.3, calculate the frictional coefficients (per molecule) for serum albumin and tropomyosin. Now, presuming that the molecular weights are correct and that the specific volume of each protein is about 0.74 cm³/g, calculate the volume of each molecule and the radius it would have if it were an unhydrated sphere. Using this radius, together with Stokes' law, you can calculate the minimum value of the frictional coefficients that these proteins could have. How do these compare with the observed values? If you assume that each molecule is a prolate ellipsoid, what axial ratios are found? Now assume that each molecule is really spherical but highly hydrated. What radius, volume, and hydration (grams of water per gram of protein) are required to account for the observed frictional coefficients? Are these values reasonable? What can one conclude?

7. Consider a "typical" globular protein, of molecular weight 40,000 g/mole, $\bar{v} = 0.75$ cm³/g, hydrated to the extent 0.3 cm³ of H_2O per cubic centimeter of dry protein.

   (a) Calculate $D_{20,w}$, assuming that $\eta = 10^{-2}$ poise (cgs), and that the protein molecule is spherical.

   (b) Repeat the calculation, assuming a prolate ellipsoid of axial ratio = 5.

   (c) Calculate the width of the boundary (as in Problem 2) after 1 h of free diffusion of the protein in part (a).

   (d) The diffusion of the protein in part (a) is studied by the laser-scattering autocorrelation method. Using $\lambda = 6328$ Å and $\theta = 90°$, what correlation time τ corresponds to a decrease of the autocorrelation function to $1/e$ of its value at $\tau = 0$? Assume that $n = 1.33$.

8. A researcher prepares a "step gradient" in a centrifuge tube by carefully layering buffer over concentrated sucrose. He does it neatly, so that the initial boundary is very sharp. He then puts the gradient away in the cold room (4°C), with the idea of using it a week later. What will the half-width of this "step" be after this time? [*Note:* The viscosity of water at 4°C is about 1.7 × the value at 20°C.]

REFERENCES

General

Cantor, C. R. and P. R. Schimmel: *Biophysical Chemistry*, W. H. Freeman and Company, Publishers, San Francisco, 1980, Chap. 10. Covers particle hydrodynamics in somewhat greater depth and detail than herein.

Diffusion

Einstein, A.: *Investigations on the Theory of the Brownian Movement*, Dover Publications, Inc., New York, 1956. A collection of Einstein's early papers on diffusion and allied topics.

Gosting, L. J.: "Measurement and Interpretation of Diffusion Coefficients of Proteins," *Adv. Protein Chem.*, **11**, 429–554 (1956). A complete and exacting discussion of free diffusion methods.

Schurr, J. M.: "Dynamic Light Scattering of Biopolymers and Biocolloids," *CRC Crit. Rev. Biochem.* (G. Fasman, ed.), **4**, 371–431 (1977). Laser-scattering methods for the determination of diffusion coefficients.

Chapter Five

SEDIMENTATION

The development of the ultracentrifuge in the 1920s marked, in one sense, the beginning of molecular biology. For the first time it was possible to measure unambiguously the molecular weights and heterogeneity of macromolecular substances. Now this instrument is used routinely for the separation and purification of macromolecular substances, the analysis of mixtures, and the determination of molecular weights and dimensions. In fact, the applications of centrifugal methods in biochemistry are so diverse that we can only touch on them here; for more detail about theory and practice, the reader should turn to the references given at the end of the chapter.

5.1 Sedimentation Velocity

We shall introduce the concept of sedimentation by a simple mechanical analysis, leaving the more mathematically elegant theory for later. Imagine a solute molecule in a solution that is held in a rapidly spinning rotor. Forgetting, for the moment, the random pushes and pulls that the molecule receives from its neighbors, we may say that there are three forces acting on it (see Figure 5.1). If the rotor turns with an angular velocity ω (radians per second),† the molecule

$\dagger \omega = (\pi/30)$ rpm, where rpm $=$ revolutions per minute.

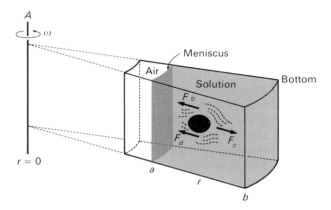

Figure 5.1 Diagram of a sedimentation experiment (not to scale). The sector-shaped cell is in a rotor spinning about the axis A at an angular velocity ω. The molecule is acted on by centrifugal, buoyant, and frictional drag forces. The cell has been given a sector shape because sedimentation proceeds along radial lines; any other shape would lead to concentration changes near the edges, with accompanying convection. The angular velocity is given by $(2\pi/60)$ times revolution per minute.

will experience a *centrifugal* force proportional to the product of its mass (m) and the distance (r) from the center of rotation: $F_c = \omega^2 rm$. At the same time, the molecule displaces some solution; this will give rise to a force equal to that which would be exerted on the mass of solution displaced: $F_b = -\omega^2 rm_0$. Finally, if as a result of these forces the molecule acquires a velocity (v) through the solution, the kind of viscous drag discussed in Chapter 4 will be experienced: There will be a *frictional force* $F_d = -fv$, where f is the frictional coefficient. As indicated in Chapter 4, this kind of situation will result in the molecule acquiring a velocity just great enough to make the total force zero:

$$F_c + F_b + F_d = 0$$
$$\omega^2 rm - \omega^2 rm_0 - fv = 0 \tag{5.1}$$

For the mass of solution displaced, m_0, we may substitute the product of the particle mass times its partial specific volume, times the solution density: $m_0 = m\bar{v}\rho$. Therefore,

$$\omega^2 rm(1 - \bar{v}\rho) - fv = 0 \tag{5.2}$$

Next, we multiply Equation (5.2) by Avogadro's number, to put things on a mole basis, and rearrange, placing molecular parameters on one side of the equation and experimentally measured ones on the other:

$$\frac{M(1 - \bar{v}\rho)}{\mathfrak{N}f} = \frac{v}{\omega^2 r} = s \tag{5.3}$$

The velocity divided by the centrifugal field strength $(\omega^2 r)$ is called the *sedimentation coefficient, s*. According to Equation (5.3), it is proportional to the molecular weight multiplied by the *buoyancy factor* $(1 - \bar{v}\rho)$ and inversely proportional to the frictional coefficient. It is experimentally determined as the ratio of velocity to field strength. The units of s are seconds. Since values of about 10^{-13} sec are commonly encountered, the quantity 1×10^{-13} sec is called 1 Svedberg. Svedberg units are conventionally denoted by the symbol S.

What happens when a centrifugal field is applied to a solution of large molecules? Figure 5.1 shows a diagram of an ultracentrifuge cell. Initially, we can imagine the molecules as being uniformly distributed throughout the solution. When the field is applied, all begin to move, and a region near the meniscus is entirely cleared of solute. Thus a *moving boundary* is formed; from the rate of motion of this boundary, we can calculate the sedimentation coefficient. Since $v = dr_b/dt$, we have, from Equation (5.3),

$$\frac{dr_b}{dt} = r_b \omega^2 s \tag{5.4}$$

Integrating gives us

$$\ln \frac{r_b(t)}{r_b(t_0)} = \omega^2 s(t - t_0) \tag{5.5}$$

where $r_b(t)$ is the position of the boundary at the time t. A graph of $\ln [r_b(t)/r_b(t_0)]$ versus $(t - t_0)$ will give $\omega^2 s$ and hence s.

If the diffusion coefficient were zero, the boundary would remain infinitely sharp as it traversed the cell. Such a situation is *approximated* by very large molecules at high rotor speeds. Figure 5.2(a) shows the sedimentation of a hemocyanin ($M = 3,850,000$). However, for lower-molecular-weight materials such as the hemocyanin subunits shown in Figure 5.2(b), diffusion appreciably spreads the boundary during sedimentation. Very small molecules, with small sedimentation coefficients and large diffusion coefficients, will not form a clear boundary at all in present ultracentrifuges. The reader should note that the photographs in Figure 5.2 are obtained with a Schlieren optical system, which records the concentration *gradient* $(\partial c/\partial r)$; an accompanying sketch shows the concentration versus distance curve. Below the boundary there is always a *plateau* region, which in most cases will extend nearly to the bottom of the cell (see also Figure 5.3). At the bottom, of course, the solute "piles up."

Having seen roughly what happens in such a *sedimentation velocity* experiment, let us now examine the process in a somewhat more precise fashion. The sedimentation process is moving toward a state of equilibrium. From the theory of irreversible processes we would expect the flow of matter to be determined by the gradient of a potential, which will become uniform at equilibrium. But which potential? Classical thermodynamics provides the answer: For a system in a field of force, the condition for equilibrium is not that the chemical potentials be uniform but that the *total potential* be everywhere constant. The total potential of a component is defined as the sum of the the chemical potential and the potential energy (per mole) of the substance in the field. Thus in a centrifugal field, where 1 mole of solute of molecular weight M has a potential energy of

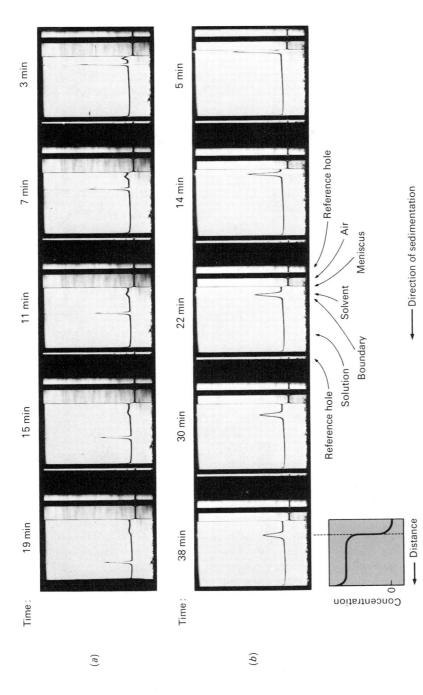

Time: 3 min   7 min   11 min   15 min   19 min

(a)

Time: 5 min   14 min   22 min   30 min   38 min

(b)

Reference hole

Reference hole

Air

Meniscus

Solvent

Boundary

Solution

— Direction of sedimentation

Concentration

0

— Distance

Figure 5.2 (*a*) Sedimentation of a hemocyanin ($M = 3,850,000$) at 42,049 rpm, 20°C. Schlieren photographs depicting the concentration gradient are shown. Times after the rotor attains speeds are given. Result: $s_{20,w} = 57.2 \times 10^{-13}$. (*b*) Like (*a*), except that these are subunits of the hemo-cyanin ($M \simeq 380,000$, $s_{20,w} = 11 \times 10^{-13}$) sedimenting at 42,040 rpm. Note that the boundary spreads more rapidly because of the greater diffusion. The small graph shows the concentration curve corresponding to the last photograph.

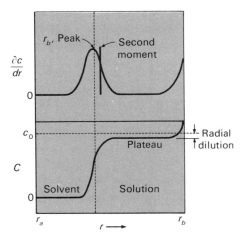

Figure 5.3 An idealized picture of a boundary, showing both the C versus r and corresponding $\partial C/\partial r$ versus r curves. Sedimentation is to the right. The second moment position is to be used for exact calculation of s. It is defined by

$$r_b^2 = \int r^2(\partial C/\partial r)dr/\int (\partial C/\partial r)\, dr,$$

where both integrals are taken across the boundary. Since r^2 occurs in the integral in the numerator, the second moment position will be further from the center of rotation than the maximum of a symmetrical boundary.

$(-\frac{1}{2}M\omega^2r^2)$, the total potential is $\bar{\mu} = \mu - \frac{1}{2}M\omega^2r^2$, and the condition for equilibrium is

$$\frac{\partial \bar{\mu}}{\partial r} = \frac{\partial \mu}{\partial r} - M\omega^2r = 0 \tag{5.6}$$

This will provide us later with a way to describe the state of *sedimentation equilibrium* in detail; for now, it allows us to write the flow equation for the solute in a two-component mixture:

$$J = -L\frac{\partial \bar{\mu}}{\partial r} = -L\left(\frac{\partial \mu}{\partial r} - M\omega^2r\right) \tag{5.7}$$

Now μ is a function of T, P, and the concentration C. Therefore,

$$\frac{\partial \mu}{\partial r} = \underset{(a)}{\left(\frac{\partial \mu}{\partial T}\right)_{P,C}\left(\frac{\partial T}{\partial r}\right)} + \underset{(b)}{\left(\frac{\partial \mu}{\partial P}\right)_{T,C}\left(\frac{\partial P}{\partial r}\right)} + \underset{(c)}{\left(\frac{\partial \mu}{\partial C}\right)_{T,P}\left(\frac{\partial C}{\partial r}\right)}$$

$$\frac{\partial \mu}{\partial r} = 0 + \bar{v}M\omega^2r\rho + \frac{RT}{C}\frac{\partial C}{\partial r} \tag{5.8}$$

The various terms have been evaluated as follows:

(a) $\partial T/\partial r = 0$, since we keep T constant.

(b) Just as $(\partial G/\partial P)_T = V$, $(\partial \mu/\partial P)_T = \bar{V}$, the partial molar volume. The partial molar volume is M times the partial specific volume.

(c) The hydrostatic pressure at a point r in the solution column is $P = P_a + \omega^2(r^2 - a^2)\rho/2$, where a is the meniscus position. This is the analog of the hydrostatic pressure in a gravitational field g: $P = P_a + g(r - a)\rho$. In both cases ρ is the *solution* density.

(d) We assume the solution to be ideal; the result follows (see Chapter 2).

Using Equation (5.8), the flow equation (5.7) becomes

$$J = L\left[\omega^2rM(1 - \bar{v}\rho) - \frac{RT}{C}\frac{\partial C}{\partial r}\right] \tag{5.9}$$

If we now express the coefficient L in terms of a frictional coefficient, $L = C/\mathfrak{N}f$, we obtain

$$J = \underline{\frac{M(1 - \bar{v}\rho)}{\mathfrak{N}f}}\omega^2 rC - \underline{\frac{RT}{\mathfrak{N}f}\frac{\partial C}{\partial r}} \qquad (5.10)$$

The underlined coefficients are identifiable as a sedimentation coefficient and diffusion coefficient, respectively:

$$J = s\omega^2 rC - D\frac{\partial C}{\partial r} \qquad (5.11)$$

$$s = \frac{M(1 - \bar{v}\rho)}{\mathfrak{N}f} \qquad (5.12)$$

$$D = \frac{RT}{\mathfrak{N}f} \qquad (5.13)$$

Neglecting certain minor corrections, such as the variation of $\bar{v}$ and ρ with pressure, Equation (5.11) provides a complete description of the processes occurring in an ultracentrifuge cell. The first and second terms on the right of Equation (5.11) clearly correspond to transport by sedimentation and diffusion, respectively. When the ultracentrifuge is started, concentration is everywhere uniform $[(\partial C/\partial r) = 0]$, and only the sedimentation term is significant. However, this transport creates a boundary; the concentration gradient therein results in diffusion, with subsequent broadening. If the field were then turned off ($\omega = 0$), the boundary would stop and spread with time, as in a simple diffusion experiment.

Equation (5.11) also provides the most rigorous definition† of s; in the plateau region of flow is given by $J_p = sC_p\omega^2 r$. It can be proved that there is a point in the boundary that moves with a velocity v' such that $J_p = C_p v'$; then $s = v'/\omega^2 r$.

This point turns out to be at the second moment of the boundary gradient curve (see Figure 5.3); for reasonably sharp boundaries one can (and almost always does) assume that this point moves with the same rate as does the point of the maximum gradient, so that only the peak of the boundary or the point of maximum slope in the C versus r curve need be followed to determine s by Equation (5.5).

Equation (5.11) can be looked at as an extension of Fick's first law to include the effect of a centrifugal field. The analog of Fick's second law can also be obtained; it requires only the application of the continuity equation [see Equation (4.10)]. However, we must be careful, for the conditions in the sector-shaped ultracentrifuge cell are a little different from those used before. Since the

---

†The necessity for a more rigorous definition than that given by Equation (5.3) should be apparent. The discerning reader will have wondered how we define, in a broad boundary, the precise *point* to use as r_b in Equation (5.3).

cross section of the cell is proportional to r, the continuity equation turns out to be†

$$\left(\frac{\partial C}{\partial t}\right)_r = -\frac{1}{r}\left(\frac{\partial rJ}{\partial r}\right)_t \tag{5.14}$$

Combining (5.14) with (5.11), we obtain the partial differential equation

$$\left(\frac{\partial C}{\partial t}\right)_r = -\frac{1}{r}\left\{\frac{\partial}{\partial r}\left[s\omega^2r^2C - Dr\left(\frac{\partial C}{\partial r}\right)_t\right]\right\}_t \tag{5.15}$$

Solutions to this rather formidable equation have been obtained for a number of specific situations; suffice it to say here that they predict the kind of behavior illustrated in Figures 5.2 and 5.3. One initially perplexing observation can easily be explained by Equation (5.15). This is that the concentration difference across a sedimenting boundary falls off with time, rather than remaining constant as in a free diffusion experiment (see Figure 5.3). If we apply Equation (5.15) to the plateau region, where $\partial C/\partial r = 0$, we obtain

$$\left(\frac{\partial C_p}{\partial t}\right)_r = -2s\omega^2 C_p \tag{5.16}$$

which can be integrated from $t = 0$ to give

$$C_p = C^0 e^{-2s\omega^2 t} \tag{5.17}$$

This *radial dilution effect* can be explained physically on the grounds that the flow out of any lamina in the plateau region is greater than the flow in, since both A and the field increase with r.

The sedimentation coefficient of a macromolecule provides a rough estimate of molecular weight. However, from Equation (5.12) it is evident that s is also influenced by molecular shape and dimensions, for the frictional coefficient enters into the denominator. A highly elongated molecule or a random coil will have a larger value of f, and hence a lower value of s, than a compact molecule of the same weight. The ambiguity introduced by the frictional coefficient can be taken care of by combining sedimentation and diffusion measurements. We must recognize that both s and D will depend on solute concentration, because of molecular interactions, If s^0, D^0, and f^0 denote quantities obtained by extrapolation to zero concentration, we have, from Equations (5.12) and (5.13),

$$s^0 = \frac{M(1 - \bar{v}\rho)}{\mathfrak{N}f^0}$$

$$D^0 = \frac{RT}{\mathfrak{N}f^0}$$

$$\frac{s^0}{D^0} = \frac{M(1 - \bar{v}\rho)}{RT} \tag{5.18}$$

---

†The reader can easily confirm this by repeating the derivation of Equation (4.10), letting the cross section A be proportional to r.

Thus a combination of s^0 and D^0 yields M; this is one of the more common ways of determining the molecular weights of macromolecules (see Table 5.1).

TABLE 5.1 *Some sedimentation and diffusion data[a]*

Substance	$s^0_{20,w}$ $\times 10^{13}$ *(sec)*	$D^0_{20,w}$ $\times 10^7$ *(cm²/sec)*	$\bar{v}_{20}$ *(cm³/g)*	$M_{s,D}$
Lipase	1.14	14.48	0.732	6,667
Lysozyme	1.91	11.20	0.703	14,400
Serum albumin	4.31	5.94	0.734	66,000
Catalase	11.3	4.10	0.730	250,000
Fibrinogen	7.9	2.02	0.706	330,000
Urease	18.6	3.46	0.730	483,000
Hemocyanin (snail)	105.8	1.04	0.727	8,950,000
Bushy stunt virus	132	1.15	0.740	10,700,000

[a]Most of these, and many other data, are listed in H. Sober (Ed.), *The Handbook of Biochemistry and Molecular Biology*, Chemical Rubber Co., Cleveland, Ohio, 1968.

In Equation (5.18) it is assumed that s and D refer to the same temperature, for the frictional coefficient depends on the viscosity, which is very temperature sensitive. Usually, both s^0 and D^0 are corrected to some standard condition. If the actual measurements were made in a buffer solution at temperature T, the values in *water at* 20°C would be

$$s_{20,w} = \frac{(1 - \bar{v}\rho)_{20,w}}{(1 - \bar{v}\rho)_{T,b}} \frac{\eta_{T,b}}{\eta_{20,w}} s_{T,b} \tag{5.19}$$

$$D_{20,w} = \frac{293.1}{T} \frac{\eta_{T,b}}{\eta_{20,w}} D_{T,b} \tag{5.20}$$

These corrections depend on the assumption that f is accurately proportional to the solvent viscosity η; if the conditions of the experiment are not too far from water at 20°C, Equations (5.19) and (5.20) are probably valid.†

A knowledge of the sedimentation coefficient *and* the molecular weight (obtained by the s, D method or some other) allows a direct calculation of the frictional coefficient. This can provide limited information about molecular shape. If we assume that the volume of the solute molecule is given by $M\bar{v}/\mathfrak{N}$, then we can calculate the radius the molecule would have were it spherical.

†It should be noted that Equations (5.19) and (5.20) correct only for variations of η, $\bar{v}$, and ρ with temperature and not for variations in macromolecular conformation per se. For example, if a particular molecule were compactly folded at 20°C and unfolded at 40°C application of these equations to data obtained at 40°C would give the hypothetical s and D that the unfolded molecule would have at 20°C.

Stokes' law [Equation (4.2)] then predicts the minimum value of the frictional coefficient (f_0) this molecule could have. If we can believe the above calculation and are willing to guess a general shape (prolate ellipsoid, oblate ellipsoid, rod, and so forth), the equations given in Table 4.1 will yield an axial ratio. But such a calculation is obviously beset with traps. Suppose the molecule is spherical but hydrated; then the frictional coefficient may be larger than f_0 just because we have calculated too small a volume by neglecting this hydration. The data appear to be really useful only when supplementary information (from viscosity measurements, for example; see Chapter 7) is available or in cases where f/f_0 is so great that we cannot reasonably attribute it to hydration. The subject will be raised again in Chapter 7.

A further limitation to attempts to deduce particle shape from sedimentation data arises from the fact that *exact* expressions for the frictional coefficient have been deduced only for simple (and often unrealistic) models—spheres, ellipsoids, and the like. There exists, however, a theory that allows one to calculate rather accurately the frictional coefficients for molecules of quite complicated shapes. In 1954, J. Kirkwood analyzed the frictional behavior of a molecule made up of a number of subunits, and showed how the hydrodynamic interaction between those subunits could be taken into account. For a particle of N identical subunits, each of frictional coefficient f_1, the frictional coefficient of the assembly can be written as

$$f_N = Nf_1 \left(1 + \frac{f_1}{6\pi\eta N} \sum_{i=1}^{N} \sum_{j=1}^{N} \frac{1}{R_{ij}} \right)^{-1} \tag{5.21}$$

Here R_{ij} is the distance between subunits i and j, and the summation is taken over all pairs. This is an exceedingly powerful equation, for in principle, it allows us to calculate the frictional coefficient of *any* object by "modeling" it in terms of a large number of small subunit beads [see Bloomfield et al. (1967)]. Such calculations are difficult, for the number of terms becomes very large. They are also inexact for very asymmetric shapes because of certain approximations in the theory. However, the Kirkwood method is quite useful and practical for the study of multi-subunit proteins. Suppose we have a protein made up of N identical subunits, each with sedimentation coefficient s_1, where

$$s_1 = \frac{M_1(1 - \bar{v}\rho)}{\mathfrak{N}f_1} \tag{5.22}$$

Then the ratio of the sedimentation coefficient of the N-mer to that of the subunit is given by the simple expression

$$\frac{s_N}{s_1} = 1 + \frac{f_1}{6\pi\eta N} \sum_{i=i}^{N} \sum_{j=i}^{N} \frac{1}{R_{ij}} \tag{5.23}$$

which follows immediately from Equations (5.21) and (5.22) and the fact that $M_N = NM_1$. The ratio s_N/s_1 will depend on the geometry of the molecule, through the subunit–subunit distances R_{ij}. If we write f_1 in Equation (5.23) as

$f_1 = 6\pi\eta R_s$, where R_s is the Stokes radius of the subunit, the result becomes even simpler:

$$\frac{s_N}{s_1} = 1 + \frac{R_s}{N}\sum_{i=i}^{N}\sum_{j=i}^{N}\frac{1}{R_{ij}} \tag{5.24}$$

The Stokes radius can, of course, be calculated from s_1 and M_1. Equation (5.24) allows us to test various postulated geometries of multi-subunit proteins, if values of s_N, s_1, M_1, and N are available.

So far, we have only alluded to the concentration dependence of s. Any interaction between the sedimenting molecules will alter the sedimentation behavior. In the most usual case the molecules may be thought of as interfering with one another, so as to make the frictional coefficient increase with concentration. If

$$f = f^0(1 + kC + \cdots) \tag{5.25}$$

where f^0 is the value of f at $C = 0$, then

$$s = \frac{s^0}{1 + kC} \tag{5.26}$$

and s will decrease with increasing C. As might be expected, this effect is most pronounced with highly extended macromolecules. Figure 5.4 shows representative data for serum albumin (a compact globular protein) and a small RNA. While both have nearly the same s^0, the concentration dependence for

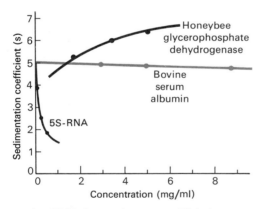

Figure 5.4 Concentration dependence of the sedimentation coefficient. Bovine serum albumin (BSA) and the ribosomal nucleic acid (5S RNA) exhibit behavior typical of compact and extended macromolecules, respectively. The behavior of honeybee glycerophosphate dehydrogenase is that to be expected for a reversibly associating substance. Data from R. L. Baldwin *Biochem. J.* **65**, 503 (1957); D. G. Comb, and Zehavi-Willmer, *J. Mol. Biol.*, **23**, 441 (1967); and R. R. Marquardt and R. W. Brosemer, *Biochem. Biophys. Acta*, **128**, 454 (1966).

the RNA is much greater. This is one reason absorption optical systems, which can detect very small concentrations of strongly absorbing substances such as nucleic acids, have come to be used so widely.

Also shown in Figure 5.4 are data from a rather different kind of system, in which s *increases* with increasing C. In all cases in which such behavior has been observed, it appears to result from a rapid monomer–polymer equilibrium. Under some conditions, such systems will yield only a single boundary, with sedimentation coefficient dependent on the proportions of monomer and polymer. Since high concentrations favor polymer formation, the s value increases with C.

If there are a number of noninteracting solute components, a sedimentation velocity experiment *may* lead to their resolution. Such a situation is depicted in Figure 5.5(*a*). If there is very little solute–solute interaction, the

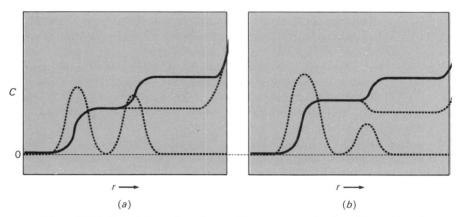

Figure 5.5 Sedimentation of a mixture of two macromolecules without (*a*) and with (*b*) the Johnston–Ogston effect. The dashed curves show the gradients that would be observed.

relative amounts of the components, as well as their individual sedimentation coefficients, can be determined. However, if the sedimentation coefficients are mutually concentration-dependent, difficulties arise in the analysis. An example is shown in Figure 5.5(*b*). Here, each component sediments more slowly in the presence of the other than it would alone. For the fast component, which is always passing through a solution of the slow component, a reduced *s* value is observed. The analysis of the composition is also more complicated. Since the slow component is retarded below the fast boundary but not above, it will tend to pile up behind the fast boundary. Most optical techniques yield only the total concentration change across each boundary, so that the composition of the system in Figure 5.5(*b*) would be misread. One would obtain a value too high for the amount of slow component and a value too low for the amount of fast component. This "Johnston–Ogston" effect can lead to very large errors in highly concentration-dependent systems.

Last, a word about homogeneity. While the observation of more than one sedimenting boundary is proof of heterogeneity, the converse is not necessarily true. Either high concentration dependence or rapid diffusion can make it impossible to resolve adjacent boundaries in the ultracentrifuge. Normally, sedimentation coefficients decrease with increasing concentration, so that molecules that diffuse to the low-concentration side of a boundary will tend to "catch up," giving an artificially sharp boundary, which can conceal heterogeneity.† Or, if the diffusion rate is too great as compared to the rate of sedi-

†Extreme examples of concentration-dependence sharpening are found with nucleic acids. Because of the great dependence of *s* on *C*, even very heterogeneous samples will give razor-sharp boundaries at high concentrations.

mentation, two gradient curves may so overlap as to yield a broad boundary with a single maximum, difficult to distinguish from that of a single component. Thus the frequently encountered statement that the "sample was found to be homogeneous in the ultracentrifuge" should be taken with a grain of salt unless rather rigorous tests have been applied to the data or unless the sample has been examined over a very wide range of experimental conditions.

5.2 Apparatus and Procedures for Sedimentation Studies

Quantitative sedimentation experiments require instruments that can provide high, controlled rotor speeds at uniform temperatures. Most modern ultra-centrifuges utilize an electric motor drive, with either mechanical or electronic automatic speed control. In addition, the rotor is usually operated in an evacu-ated chamber to reduce heating by air friction, and refrigeration and temperature control are generally provided. Two types of ultracentrifuges, *analytical* and *preparative*, are distinguished by whether or not an optical system for the observation of the cell is provided. Some commercially available analytical ultracentrifuges can attain rotor speeds as high as 60,000 rpm, giving a field at the cell poistion (about 6.8 cm from the center of rotation) of nearly 280,000 times gravity.

Optical systems for analytical ultracentrifuges are of three general types. The *Schlieren* system operates on the principle that light passing through a refractive index gradient is deflected, as if it were passed through a prism. The magnitude of the deflection is proportional to the steepness of the gradient. The optical system is contrived so as to produce a distored image of a wire on a photographic plate. The amount by which the wire image is displaced at any point in the image of the ultracentrifuge cell measures the refractive index gradient $(\partial n/\partial r)$ at that point. Since the refractive index increases linearly with solute concentration, this displacement will also be proportional to $(\partial C/\partial r)$. Thus images like those in Figure 5.2 are obtained. The point at which the peak has a maximum is the point of steepest concentration gradient in the boundary. The system is rarely used at the present time, but the reader will find many such photographs in the older literature.

Most contemporary sedimentation velocity experiments use a *scanning absorption optical system*. The ultracentrifuge is equipped with a light source and monochromater, so that a wavelength can be selected at which the solute absorbs maximally. The ultracentrifuge cell is illuminated by this light each time it revolves through the optical path. Lenses or focusing mirrors are arranged to form a cell image, as shown in Figure 5.6(*a*). This image is then scanned by a photomultiplier behind a narrow slit, which travels across the cell image in the *r* direction. Usually, a double-sector cell is employed, so that the photo-multiplier alternately observes solution and solvent; thus the whole system acts as a double-beam spectrophotometer. Ar each point *r*, the absorbance difference

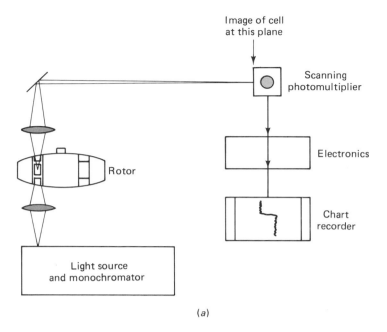

(a)

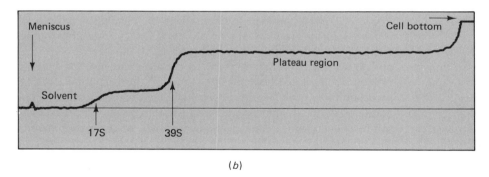

(b)

Figure 5.6 (*a*) Photoelectric scanning optical system for an analytical ultracentrifuge. (*b*) A typical scan obtained in a sedimentation velocity experiment. The slow reassociation of 17S hemocyanin monomers into 39S tetramers was being studied. Both boundaries are clearly resolved. The monochromator was set at 345 mm, in a strong absorption band of hemocyanin. Data from F. Arisaka, Ph.D. thesis, Oregon State University, 1977.

between the solution sector and the reference sector is recorded. The kind of output obtained is shown in Figure 5.6(*b*); it is a graph of absorbance versus *r* at a particular time in the sedimentation experiment. The boundary position is usually defined as the midpoint of the absorbance step and *s* can then be calculated from Equation (5.5).

The scanning absorption system possesses a number of advantages over earlier optical techniques. Since the wavelength can be varied, substances that strongly absorb at particular wavelengths can be studied even in the presence of contaminating materials. The high absorbance of nucleic acids makes it possible to do experiments with quite small amounts of material; a moving-boundary sedimentation experiment can be done easily with as little as 10 μg of DNA. Finally, the electronic output of the scanner system can be digitized for on-line computer processing and analysis of the data.

The development of interferometric optical systems for the study of transport processes has increased the accuracy of some measurements by nearly one order of magnitude. Most commonly employed is the Rayleigh interferometer shown in Figure 5.7(a). The source slit is parallel to the cell axis. A double cell is used, with solvent in one side and the solution-solvent boundary in the other. If a pair of slits are inserted, one behind each channel, a double-slit interference

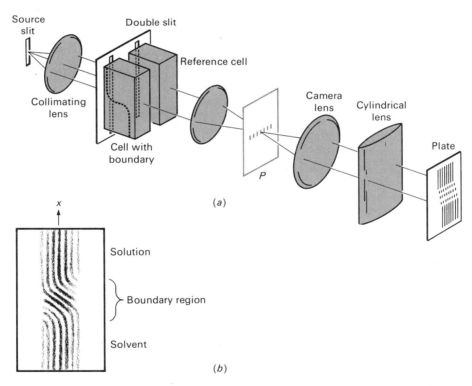

(a)

(b)

Figure 5.7 (*a*) A schematic diagram of the Rayleigh interference system. Each layer in the cell can be thought of as providing its own interference pattern. These are all superimposed at *P*, but are sorted out by the astigmatic optical system. The image shown on the plate is intended to illustrate this idea. Actually, the pattern will be similar to that shown (*b*). (*b*) The kind of fringe pattern produced by a boundary in the Rayleigh system. Only the central diffraction band of the image shows up; superimposed on this are the Rayleigh fringes.

pattern might be expected at the focal plane *P*. However, the actual pattern at this plane will be much more complex, because of the bending of light and because the optical path difference is different at various heights in the double cell. To sort this out, the same kind of astigmatic system employed in the Schlieren system is used. The cell is in focus at the plate *along the cell axis*; but at right angles to this, because of the cylindrical lens, the focal plane *P* is in focus. What happens can best be imagined by visualizing the cell as divided into a series of layers, perpendicular to its axis. Each layer has "its own" interference pattern, with fringes shifted according to the refractive index difference between the two channels at that layer. Each layer is, of course, focused at the correct height. The overall pattern obtained is shown in Figure 5.7(*b*). The progressive shifting of the fringes as they pass through the boundary region means that each fringe traces the refractive index versus distance curve. The refractive index difference (Δn) between two points on the cell axis can be judged by the number of fringes crossed (Δj) in going between these two points:

$$\Delta j = \frac{a\,\Delta n}{\lambda} \tag{5.27}$$

Here *a* is the cell thickness and λ the wavelength of the light used. This makes measurement of refractive index (and hence concentration) difference very easy and accurate.

Since sedimentation velocity experiments are so widely used in biochemistry, a few words about proper procedure are in order. A careful study will include measurements at several concentrations; these should extend to as low a concentration as the optical system and substance will allow. For biopolymers, a buffer solution, usually with ionic strength at least 0.1, should be used. This is because charged polymers may be retarded in sedimentation by an effect analogous to the Donnan effect in osmometry (see Chapter 2). Temperatures should be controlled precisely, and the rotor speed should be actually measured during the experiment. For the most accurate work, especially with broad boundaries, the second moment should be used for the boundary; for routine studies the maximum in the gradient can be employed.

With proper attention to such details, sedimentation coefficients can be determined with precision of better than 0.5 percent. Since *D* can also be quite accurately measured, precise molecular weights should be obtainable from Equation (5.18). Actually, the quantity hardest to obtain with high accuracy is the partial specific volume. Since $\bar{v} \simeq 0.7$ for most proteins, an error of 1 percent in $\bar{v}$ gives an error of about 3 percent in $(1 - \bar{v}\rho)$ and, consequently, in *M*.

5.3 Sedimentation Equilibrium

Now let us examine that state of equilibrium toward which a sedimentation experiment is moving. We shall be thinking of an experiment in which the rotor speed is not as great as in a velocity experiment; we do not want all of the solute to be packed in the bottom of the cell. To obtain equations describing the gradient that will exist at equilibrium, we might start with the general equilibrium condition [Equation (5.6)] without ever introducing the concept of flow. However, to save a lot of duplication in the derivations, let us take the following point of view: When equilibrium is finally established, flow must vanish at all points in the ultracentrifuge cell. There will still be the random motion of molecules, but no net transport of matter will henceforth occur as long as the ultracentrifuge continues to operate at the same rotor speed and temperature. Thus we must set $J = 0$ in Equation (5.9):

$$L\left[\omega^2 r M(1 - \bar{v}\rho) - \frac{RT}{C}\frac{dC}{dr}\right] = 0 \qquad (5.9)$$

$$\frac{1}{C}\frac{dC}{dr} = \frac{\omega^2 r M(1 - \bar{v}\rho)}{RT} \qquad (5.28)$$

We have replaced the partial derivative of Equation (5.9) with a total derivative, since C is no longer a function of time but only of r.

Equation (5.28) describes the concentration gradient at equilibrium for a single solute component in an ideal two-component solution. By integrating it between the meniscus (a) and some point r, we can see that C depends on r in an exponential fashion:

$$C(r) = C(a)e^{\omega^2 M(1-\bar{v}\rho)(r^2-a^2)/2RT}$$

or

$$\ln\frac{C(r)}{C(a)} = \frac{\omega^2 M(1 - \bar{v}\rho)(r^2 - a^2)}{2RT} \qquad (5.29)$$

Alternatively, we may write this as

$$\frac{d\ln C}{d(r^2)} = \frac{\omega^2 M(1 - \bar{v}\rho)}{2RT} \qquad (5.30)$$

Obviously, a graph of $\ln C$ versus r^2 should give a straight line, with a slope of $\omega^2 M(1 - \bar{v}\rho)/2RT$ (see Figure 5.8). From this, M may be determined, provided that $\bar{v}$ has been measured (or estimated) with sufficient accuracy. The method is thermodynamically rigorous and involves no assumptions about particle shape or hydration.

At the present time, the concentrations C at different points in the cell are usually determined in one of two ways. Easiest, but less accurate, is to use the absorption optical system (Section 5.2). A more accurate method utilizes the interferometric optical system, as shown in Figure 5.8. The interference fringes can be measured with great precision. There is only one difficulty; the number of fringes crossed between two points in the cell measures the *difference* in concentration between these two points; the *absolute* concentration is not given. This problem is usually circumvented as shown in Figure 5.8(*a*). The centrifuge is run at a high enough speed that the solute concentration becomes essentially zero at the meniscus. Then the number of fringes crossed in going from the meniscus to any point r does give a measure of the absolute concentration at r.

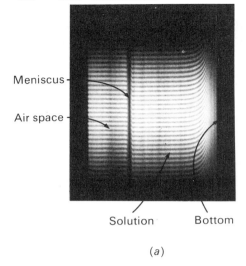

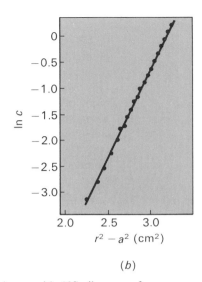

Meniscus

Air space

Solution   Bottom

(a)

(b)

Figure 5.8 A sedimentation equilibrium experiment with 30S ribosomes from *Escherichia coli*. The interferogram is shown in (a), while (b) depicts a graph according to Equation (5.29). Conditions: $C^0 = 0.35$ mg/ml; $T = 4°C$.

This is usually referred to as the *meniscus depletion* or *Yphantis method,* after its discoverer, D. Yphantis.

Since the sedimentation equilibrium method is based on equations applicable to thermodynamic equilibrium, it is easy to extend it in a rigorous way to include such effects as solution nonideality. For example, it is easy to show that for a nonideal solute,

$$\frac{d \ln C}{d(r^2)} = \frac{\omega^2 M(1 - \bar{v}\rho)}{2RT[1 + C(d \ln y/dC)]} \tag{5.31}$$

If $(d \ln y/dC)$ is positive, as with a positive second virial coefficient, Equation (5.31) indicates that a graph of ln C versus r^2 will not be linear, but will show downward curvature at higher C. The true molecular weight is then obtained by extrapolating the slope to $C = 0$.

So far we have considered only the equations applicable to a homogeneous solute. If the solute is actually a mixture of molecular species, as for example a heterogeneous high polymer or a protein that can associate to different states of aggregation, the analysis becomes a bit more complex. It is not very difficult, however, if (as is likely in the cases above) all of the components of the mixture have the same value of $\bar{v}$. We may then write Equation (5.28) for each solute species as

$$\frac{1}{C_i}\frac{dC_i}{dr} = \frac{\omega^2 r(1 - \bar{v}\rho)}{RT}M_i$$

or

$$\frac{dC_i}{d(r^2)} = \frac{\omega^2(1 - \bar{v}\rho)}{2RT}M_i C_i \tag{5.32}$$

Summing over all n solute species, we obtain

$$\sum_1^n \frac{dC_i}{d(r^2)} = \frac{dC}{d(r^2)} = \frac{\omega^2(1 - \bar{v}\rho)}{2RT} \sum_1^n C_i M_i \qquad (5.33)$$

since

$$C = \sum_1^n C_i \qquad (5.34)$$

Dividing Equation (5.33) by (5.34), we find that

$$\frac{1}{C} \frac{dC}{d(r^2)} = \frac{\omega^2(1 - \bar{v}\rho)}{2RT} \frac{\sum_1^n C_i M_i}{\sum_1^n C_i}$$

or

$$\frac{d \ln C}{d(r^2)} = \frac{\omega^2(1 - \bar{v}\rho)}{2RT} M_{wr} \qquad (5.35)$$

Here we have made use of the definition of M_w from Table 2.3. We call it M_{wr} to call attention to the fact that this is the weight-average molecular weight of the mixture that exists at the point r in the cell. Since the different components will distribute differently in the cell, with higher-molecular-weight materials preferentially near the cell bottom, M_{wr} will increase with r. Thus, as Equation (5.35) shows, a heterogeneous mixture will yield an upwardly curving $\ln C$ versus r^2 graph. Sedimentation equilibrium can thus be diagnostic for solute homogeneity or heterogeneity.

One other advantage of sedimentation equilibrium as a method for molecular weight determination should be mentioned. Not only is it rigorous and accurate, but it can be used over a remarkably wide molecular weight range. The rotor speed can be adjusted so as to yield usable concentration gradients for molecules ranging from a few hundreds to tens of millions. As is illustrated in Table 5.2, results of high precision can be obtained.

In examining Table 5.2, the discerning reader may wonder what the present utility of sedimentation equilibrium experiments may be. Clearly, sequencing a polypeptide chain gives a much more accurate—indeed, exact—answer. Approximate results can be obtained more quickly and cheaply by SDS-gel electrophoresis (see Chapter 6). But either of these methods yields only the mass of

TABLE 5.2 *Some sedimentation equilibrium results*

Substance	*Molecular Weight from Chemical Formula*	*Molecular Weight from Sedimentation Equibrium*
Sucrose	342.3	341.5
Ribonuclease	13,683	13,740
Lysozyme	14,305	14,500
Chymotrypsinogen *A*	25,767	25,670
30S *E. coli* ribosomes	—	900,000
Whelk hemocyanin	—	13,200,000

the *individual* polypeptide chains. Most proteins exist in their functional forms as multichain structures. The determination of the molecular weights of such aggregates by a physical technique such as sedimentation equilibrium allows the determination of the number of subunits involved in the complex.

A complementary use is in the analysis of association–dissociation equilibria of macromolecules. If a system in chemical equilibrium is centrifuged to a state of sedimentation equilibrium, the mixture of species present at each point in the ultracentrifuge cell will depend on the total concentration at that point. For example, if a monomer-dimer equilibrium exists, the fraction of monomer present at each point must obey Equation (3.93). Thus the equilibrium is "sampled" at a series of different concentrations in a single experiment. It will be noted that Equation (5.35) provides a means to calculate the weight-average molecular weight M_{wr} at each point. This is associated with a total concentration $C(r)$ at this point. By definition, M_{wr} may be written in this case as

$$M_{wr} = \frac{C_M M_M + C_D M_D}{C_M + C_D} = f_M M_M + f_D M_D \qquad (5.36)$$

where M_M and M_D are the molecular weights of monomer and dimer, and f_M and f_D are the weight fractions of monomer and dimer at point r. Then, since $f_D = 1 - f_M$ and $M_D = 2M_M$, we can rearrange Equation (5.36) to yield

$$f_M = 2 - \frac{M_{wr}}{M_M} \qquad (5.37)$$

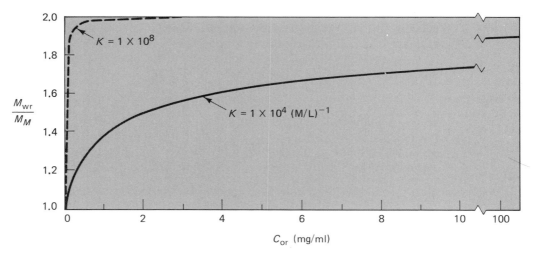

Figure 5.9 A graph of M_{wr} versus $C(r)$ as might be obtained for a reversible momomer–dimer reaction observed in a sedimentation equilibrium experiment. Here $C(r)$ is the total solute concentration at point r. The reaction involves a monomer of molecular weight 20,000, with association constants of 1×10^8 moles/liter and 1×10^4 moles/liter. Note that the former would be almost impossible to study by this method.

Thus we can obtain the composition of the mixture at each point in the cell and calculate an apparent equilibrium constant at each point. Constancy of this value is a test that this is a monomer–dimer equilibrium. This can also be chekced by the behavior of the M_{wr} versus $C(r)$ curve, as shown in Figure 5.9. As $C(r) \rightarrow 0$, $M_{wr} \rightarrow M_1$, as $C(r) \rightarrow \infty$, $M_{wr} \rightarrow 2M_1 = M_D$. Other, more complex association schemes can be tested for by more sophisticated methods. Sedimentation equilibrium has become one of the most versatile techniques for the study of macromolecular association processes.

5.4 Density Gradient Sedimentation

If sedimentation of a macromolecule is carried out in the presence of a density gradient that has been produced by a gradient in the concentration of some "inert" substance, the resolving power of the ultracentrifuge can be greatly enhanced. There are two kinds of techniques: In the equilibrium method, the concentration gradient in the inert substance is obtained by running the ultracentrifuge until equilibrium is attained, and the macromolecular component comes to equilibrium in this gradient. Alternatively, one may produce a density gradient with a substance such as sucrose by carefully layering solutions of decreasing density and then studying the sedimentation transport of a macromolecular material through this gradient.

Let us consider the latter technique first. The sequence of operations as usually performed is shown in Figure 5.10. The sucrose gradient produced in this way is, of course, not stable; it will eventually disappear by diffusion. But this process is slow, and experiments of several hours' duration can be performed without too much change occurring in the gradient. The macromolecular solution can be layered onto such a gradient, providing that it is quite dilute and less dense than the top sucrose solution.

The advantage in resolving power of the gradient technique over the conventional sedimentation transport experiment is obvious. In the conventional method, only the most slowly sedimenting material can be recovered in pure form, whereas the density gradient method gives a *spectrum* of the mixture. The crude-appearing sampling technique actually works well and little remixing is produced. More sophisticated and reproducible techniques have been developed for emptying the tubes, when high precision is needed. From the tube number in which a given component appears, the distance traveled can be estimated and s can be calculated. However, the variation in both density and viscosity in the sucrose gradient makes the velocity vary during sedimentation and complicates the calculations. While the inclusion of marker substances of known sedimentation coefficient increases the accuracy, the method still cannot give the precision of the conventional techniques. Density gradient centrifugation does, however, have the enormous advantage that a wide variety of assay techniques can be used to detect the various components. Materials can

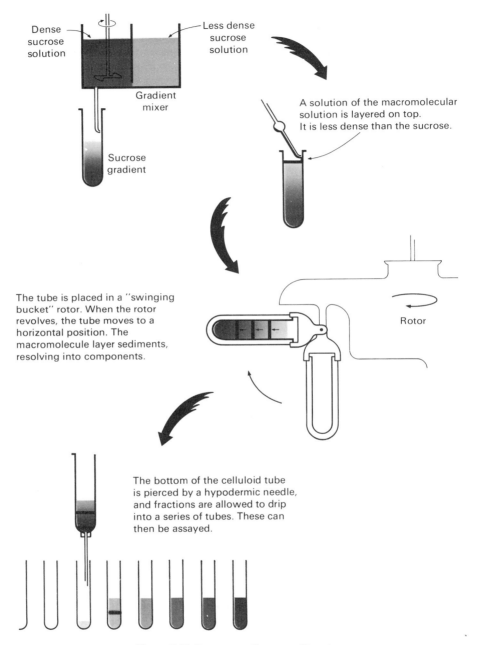

Dense sucrose solution

Less dense sucrose solution

Gradient mixer

Sucrose gradient

A solution of the macromolecular solution is layered on top. It is less dense than the sucrose.

The tube is placed in a "swinging bucket" rotor. When the rotor revolves, the tube moves to a horizontal position. The macromolecule layer sediments, resolving into components.

Rotor

The bottom of the celluloid tube is pierced by a hypodermic needle, and fractions are allowed to drip into a series of tubes. These can then be assayed.

Figure 5.10 Sucrose gradient centrifugation.

be specifically labeled with radioactive isotopes or discriminating enzymatic assays can be used. In this way the sedimentation of a minor component in a crude mixture can be followed, a feat impossible with the optical methods used in most analytical ultracentrifuges.† Figure 5.11 shows the sedimentation of an enzyme mixture in a sucrose gradient.

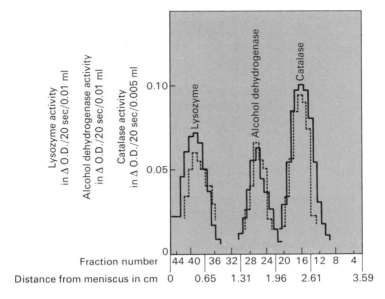

Figure 5.11 A sucrose gradient sedimentation velocity experiment with mixtures of three enzymes. The solid lines show enzyme activities in a mixture of the three enzymes in tris buffer. The dashed lines show activities in a mixture prepared using a crude bacterial extract as the solution medium. In each case, sedimentation was for 12.8 hr at 3°C, 37,700 rpm. From R. B. Martin and B. Ames, *J. Biol. Chem.*, **236**, 1372 (1961).

The density gradient sedimentation equilibrium experiment works on an entirely different principle. Suppose, as a specific example, that a solution containing a small amount of high-molecular-weight nucleic acid and a high concentration of a dense salt such as cesium chloride is placed in an ultracentrifuge cell and spun for a long time at high speed. The salt will eventually reach sedimentation equilibrium, giving a concentration gradient like that described by Equation (5.29). (Actually, this equation will not be obeyed exactly, for the concentrated solution will be nonideal.) This will yield a density

†A variant in the density gradient velocity experiment has been devised that allows use of the analytical ultracentrifuge. It utilizes a special cell in which a layer of the solution is automatically layered on a salt solution as the ultracentrifuge is accelerated. Diffusion at the edge of the layer produces a density gradient that stabilizes the zone. The sedimentation of bands can then be observed with the usual optical system.

gradient in the cell. The experiment becomes interesting if the salt concentration has been so chosen that at some point (r_0) in the cell, $\rho(r_0) = 1/\bar{v}$, where $\bar{v}$ is the specific volume of the macromolecule. The quantity $(1 - \bar{v}\rho)$ will then be positive above this point and negative below it, so we would expect sedimentation of the macromolecule to converge to the point r_0. Eventually, equilibrium will be established between the tendency for the macromolecules to diffuse away from r_0 and the effect of sedimentation to drive them back.

We can use Equation (5.28) to describe the sedimentation equilibrium of the macromolecule provided we remember that ρ depends on r. For the limited region about r_0 in which we are interested, we may assume that the density gradient is linear:

$$\rho(r) = \frac{1}{\bar{v}} + (r - r_0)\frac{d\rho}{dr} \tag{5.38}$$

where the value of $d\rho/dr$ is determined by the nature of the salt and its concentration, as well as the rotor speed. From Equation (5.28), then,

$$\frac{1}{C}\frac{dC}{dr} = \frac{\omega^2 r M}{RT}\left\{1 - \bar{v}\left[\frac{1}{\bar{v}} + (r - r_0)\frac{d\rho}{dr}\right]\right\}$$

$$= -\frac{\omega^2 r M}{RT}\bar{v}(r - r_0)\frac{d\rho}{dr} \tag{5.39}$$

Recognizing that $r \simeq r_0$ and $(r - r_0)\,dr = \frac{1}{2}\,d(r - r_0)^2$, this becomes

$$d \ln C \simeq -\frac{\omega^2 r_0 M}{2RT}\bar{v}\frac{d\rho}{dr}d(r - r_0)^2 \tag{5.40}$$

or, upon integration,

$$C(r) = C(r_0) \exp\left[-\frac{\omega^2 r_0 M \bar{v}}{2RT}\frac{d\rho}{dr}(r - r_0)^2\right] \tag{5.41}$$

Thus at equilibrium the macromolecular component will be distributed in a Gaussian band about r_0. The breadth of this band will depend on $M^{-1/2}$; thus a substance of infinite M would be packed into an infinitely thin band. The measurement of molecular weight is complicated by the disastrous effects of even minor density heterogeneity. Obviously, a narrow distribution of densities will produce a broadened band that will be difficult to distinguish from the Gaussian band for a lower-weight substance.

While the method can be used for molecular weight determination, its greatest utility lies in the sensitivity of the band position to solute density. Figure 5.12 shows the clear resolution of nucleic acids differing only in that one contains ^{14}N and the other ^{15}N.

The experiment depicted in Figure 5.12 was of great importance in the development of molecular biology. By devising a technique that allowed the resolution of ^{14}N and ^{15}N nucleic acids, Meselson et al. were able to domenstrate, for the first time, that the replication of DNA was a semiconservative process. It was this same density gradient sedimentation equilibrium technique that led

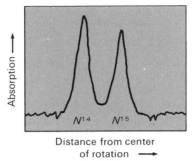

Figure 5.12 Density gradient sedimentation equilibrium of a mixture of DNA's from bacteria grown on ^{15}N and ^{14}N media. Note that separation of the bands is excellent, even though the densities differ only from the isotope substitution. From M. Meselson, F. W. Stahl, and J. Vinograd, *Proc. Natl. Acad. Sci. USA*, **43**, 581 (1957).

to the discovery of "satellite" DNAs. The density at wich a DNA molecule bands in such a gradient is a function of the $(G + C)/(A + T)$ ratio. Certain repetitive DNA sequences, which are particularly rich in either G/C or A/T base pairs, will form "satellite bands" to the "main band" DNA. Their discovery gave the first solid indication of the complexity of the eukaryotic genome. The density gradient sedimentation experiment can be used as a preparative method for such DNA, and to separate DNA from RNA and DNA/RNA hybrids. A variety of high-density salts are used to prepare gradients which are best suited for different isolations or analyses.

Our treatment of the theory of density gradient sedimentation equilibrium has been somewhat unsophisticated, since we have neglected solute–solute and solute–solvent interactions. A more exact discussion, in which the nonideality is considered, yields the important difference that in this case the macromolecular weight and specific volume refer to the *solvated* molecule. We have erred in treating the system as one containing only two components. Since a high concentration of the salt is required, a full-fledged three-component treatment is needed for accurate interpretation.

PROBLEMS

1. The best data for bovine serum albumin at 25.00°C are

$$s^0 = 5.01 \times 10^{-13} \text{ sec}$$

$$D^0 = 6.97 \times 10^{-7} \text{ cm}^2/\text{sec}$$

$$\bar{v} = 0.734 \text{ cm}^3/\text{g}$$

Calculate $M, f/f_0$, and the axial ratio, assuming that the molecule is an unhydrated prolate ellipsoid. Alternatively, assuming the molecule to be spherical, what hydration (cm³ H_2O/cm³ protein) is required to account for f/f_0? At 25°C, the viscosity of water is 0.895×10^{-2} (cgs units).

2. A sedimentation velocity experiment was performed with subunits of aldolase at 3°C, with a rotor speed of 51,970 rpm. Data for the boundary position versus time are given below. The time of the first photograph has been taken as zero.

r_b (cm)	$t - t_0$ (sec)
6.2438	0
6.3039	3,360
6.4061	7,200
6.4920	11,040
6.5898	14,840
6.6777	18,720

Calculate the sedimentation coefficient. Correct to $s_{20,w}$, assuming that $\bar{v} = 0.73$ at 3°C and 0.74 at 20°C. The viscosity and density of the buffer, relative to water were 1.016 and 1.003, independent of temperature. Other data can be found in standard handbooks.

*3.   For a heterogeneous material, we can define a weight-average sedimentation coefficient and a weight-average diffusion coefficient by

$$\bar{s}_w = \frac{\sum s_i C_i}{\sum C_i} \qquad \bar{D}_w = \frac{\sum D_i C_i}{\sum C_i}$$

Show that for a polymer for which the frictional coefficient is proportional to some power (b) of the molecular weight ($f_i = KM_i^b$), the M calculated from s_w and D_w is *not* in general a weight-average molecular weight. What average will be obtained if $b = 1$? If $b = 0$?

4.   For a homologous series of random-coil molecules, one might expect the frictional coefficient to be proportional to the radius of gyration. We observe that such series (that is, polymers of a given monomer) obey equations such as

$$s = KM^a$$

What would you predict the value of a to be? For DNA, a is found to be about 0.33. How can this result be explained?

5.   Nucleosomes are protein–DNA complexes which are the subunits of chromatin. The following data have been obtained:

   (*a*)   Sedimentation equilibrium experiment at 5621 rpm, $T = 8.8$°C:

r (cm)	*Absorbance*
6.960	0.200
7.020	0.300
7.058	0.400
7.087	0.500
7.110	0.600
7.130	0.700
7.149	0.800
7.166	0.900

     Assuming that $\bar{v} = 0.67$, calculate the molecular weight. Does the preparation appear to be homogeneous?

   (*b*)   From sedimentation velocity experiments, we obtain $s_{20,w}^0 = 11.4$S. From this, and the data given above, calculate the Stokes' radius and the quantity f/f_0. Describe nucleosomes from these data.

(*c*)   The nucleosome contains about 150 base pairs of DNA. Can this DNA be extended in the particle?

6.   Using the data in Problem 1, calculate the position and width at half-height of a serum albumin boundary after 1 h of sedimentation at 50,000 rpm. Assume that the meniscus is at 6.00 cm from the center of rotation.

7.   (*a*)   A protein with $\bar{v} = 0.72$ cm^3/g is studied by sucrose gradient centrifugation at 5°C. If the gradient runs from 10 to 30 percent sucrose, by what percent will the sedimentation coefficient decrease as the protein proceeds from the meniscus to the bottom of the tube? The following data for sucrose solution at 5°C are necessary:

Percent Sucrose	ρ (g/cm^3)	η (centipoise)
10	1.0406	2.073
30	1.1315	4.422

(*b*)   If the meniscus is 8.0 cm, and the bottom of the tube 16.0 cm from the center of rotation, by what percent will the *velocity* of sedimentation decrease in traversing the tube?

8.   Polyribosomes have much larger sedimentation coefficients than do individual ribosomes or their subunits. Devise an experiment, using sucrose gradient centrifugation, to show that only *poly*ribosomes are active in protein synthesis.

9.   The bouyant densities of *E. coli* DNA and salmon sperm DNA in CsCl are 1.712 and 1.705 g/cm^3, respectively. Suppose that a mixture of these two DNAs, each of $M = 2 \times 10^6$ g/mole, is studied by density gradient sedimentation equilibrium. The rotor speed is 45,000 rpm, $T = 25$°C, and the density gradient in the region of banding (about 7 cm from the center of rotation) is 0.1 g/cm^4. Compare the peak separation to the band width (as measured by the standard deviation, σ) under these conditions. Is resolution possible? Would the resolution be improved by dropping the rotor speed to 30,000 rpm? You may assume that the density gradient is proportional to ω^2.

10.   (*a*)   The volume of an anhydrous spherical molecule is given by $M\bar{v}/\mathfrak{N}$. Using this fact, show that such molecules should obey the equation

$$\log \frac{s\bar{v}^{1/3}}{1 - \bar{v}\rho} = \frac{2}{3} \log M + \text{constant}$$

and evaluate the constant at 20°C, in water.

(*b*)   Graph the data given in Table 5.1 according to this relationship. What conclusions can you draw? All of the molecules listed, except fibrinogen, are considered to be "globular" proteins.

11.   Using the Kirkwood theory:

(*a*)   Predict the sedimentation coefficient of a dimer made from two spherical subunits in contact, if each subunit has $s_1 = 4.5$S.

(*b*)   Calculate the ratios s_4/s_1 for linear and tetrahedral tetramers made from contacting spheres.

REFERENCES

Bloomfield, V., W. O. Dalton, and K. E. Van Holde: *Biopolymers*, **5**, 135–148 (1967). Application of the Krkwood theory to calculation of frictional coefficients for molecules of various shapes.

Fujita, H.: *Foundations of Ultracentrifugal Analysis*, John Wiley & Sons, Inc., New York, 1975. The ultimate reference for questions on ultracentrifugation theory.

Meselson, M. and F. W. Stahl: *Proc. Natl. Acad. Sci. USA*, **44**, 671–682 (1958). An elegant use of the density gradient method.

Nichol, L. W. and D. W. Winzor: *Migration of Interacting Systems*, Clarendon Press, Oxford, 1972. A brief description of theory and methods for studying interacting macromolecules.

Sheller, P.: *Centrifugation in Biology and Medical Science*, John Wiley & Sons, Inc., New York, 1981. Entirely devoted to density gradient sedimentation; many practical hints.

Svedberg, T. and K. O. Pedersen: *The Ultracentrifuge*, Oxford University Press, London, 1940. The classical reference to ultracentrifugation; now somewhat dated but of historical interest.

Van Holde, K. E.: "Sedimentation Analysis of Proteins," in *The Proteins*, 3rd ed. (H. Neurath and R. L. Hill, eds.), Academic Press, Inc., New York, 1975, Vol. I, pp. 225–291. An overview of methods for studying proteins.

Yphantis, D. A.: *Biochemistry*, **3**, 297–317 (1964). The original paper on the meniscus-depletion method.

Chapter Six

ELECTRIC FIELDS

The great majority of the polymers of biological interest are electrically charged. Like low-molecular-weight electrolytes, polyelectrolytes are somewhat arbitrarily classified as "strong" or "weak" depending on the ionization constants of the acidic or basic groups. They may be strong polyacids, such as the nucleic acids; weak polybases, such as poly-*l*-lysine; or polyampholytes, such as the proteins. As shown in Chapter 2, this polyelectrolyte character has a considerable influence on the solution behavior of these substances. It is not surprising, then, that there has been a great deal of interest in methods for the determination of the charge carried by biopolymers. Conversely, the fact that otherwise similar molecules may sometimes differ in charge has been widely used as the basis for electrical separation methods. Finally, the fact that the distribution of charge on a macromolecule may be asymmetric (even when the net charge is zero) has been utilized to orient macromolecules in electric fields.

It should be recalled that there are two basic ways in which a molecule can respond to an applied electrical field. If it carries a net charge, it will *migrate* in the field; this is the basis of all studies of electrophoresis. In addition, the molecule may either have an asymmetric distribution of charge or be polarizable. In such case, an electric field will produce some degree of *orientation*. It is the purpose of this chapter to examine these two kinds of responses to electrical fields and to show how they can be used.

6.1 Electrophoresis: General Principles

The transport of particles by an electrical field is termed electrophoresis. Formally, the process is very like sedimentation, for in both cases a field, which is the gradient of a potential, produces the transport of matter. However, since electrophoresis will depend on the charge on a macromolecule rather than its mass, it provides an additional "handle" for the analysis and separation of mixtures.

We might hope to use the theory of irreversible processes to develop a very general theory of electrophoretic transport (and, in fact, some progress has been made in this direction) but certain complications in the analysis of real systems make a mechanistic point of view more fruitful for our purposes. These difficulties arise mainly because electrophoresis is almost always carried out in aqueous solutions containing buffer ions and salts in addition to the macromolecule of interest; at the very least there must be present the counterions that have dissociated from the macromolecule to provide its charge. This means that we study the molecule not alone but in the presence of many other charged particles, and these will both influence the local field and interact with the macromolecule so as to make analysis difficult. To begin, then, let us consider an unreal simplified situation, an isolated charged particle in a nonconducting medium [Figure 6.1(*a*)]. We may write the basic equation for electrophoresis in exactly the same fashion as was used in sedimentation; we assume that the force on the particle, when suddenly applied, causes an immediate acceleration to a velocity at which the viscous resistance of the medium just balances the driving force. The force (in dynes) experienced by a particle in an electrical field is given

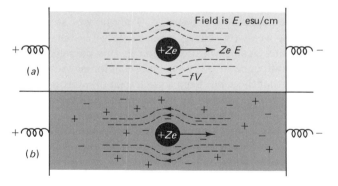

Figure 6.1 (*a*) An idealized model in which a particle of charge *Ze* is placed in an electric field in a nonconducting solvent. (*b*) A more realistic model, in which a charged macromolecule is subjected to an electric field in an aqueous salt solution. The small ions form an ion atmosphere around the macromolecule in which ions of opposite charge to that of the macromolecule predominate. This ion atmosphere is distorted by the field and by the motion of the macromolecule.

by Coulomb's law:

$$F = zeE \tag{6.1}$$

where z is the number of electron units of charge ($e = 4.8 \times 10^{-10}$ esu)† and E is the field in electrostatic units of potential per centimeter. Then the velocity (v) is given by

$$fv = zeE \tag{6.2}$$

where f is a frictional factor, which, in this case at least, we can identify with the factor that would be determined by a sedimentation or diffusion experiment with the same particle. Equation (6.2) can be rearranged to define an *elctrophoretic mobility*, a quantity very analogous to the sedimentation coefficient:

$$U = \frac{v}{E} = \frac{ze}{f} \tag{6.3}$$

If the particle happens to be spherical, Stokes' law applies and one may write

$$U = \frac{ze}{6\pi\eta R} \tag{6.4}$$

where R is the particle radius and η the solvent viscosity. In principle this equation could be generalized to take into account hydration and deviations from spherical shape, as was done in the cases of sedimentation and diffusion. However, in dealing with real macroions in aqueous solution, there are more formidable complications to consider first. In such cn environment, the macromolecule is surrounded by an *ion atmosphere*, a region in which there will be a statistical preference for ions of opposite sign [see Figure 6.1(b)]. This means that the field experienced by the macromolecule will differ appreciably from that which we measure as being imposed by the electrodes. To a degree, correction for this effect can be made if one uses the Debye–Huckel theory to calculate the properties of this ion atmosphere. A rough approximation gives

$$U = \frac{ze}{6\pi\eta R} \frac{X(\kappa R)}{1 + \kappa R} \tag{6.5}$$

where

$$\kappa = \left(\frac{8\pi\mathfrak{N}e^2}{1000DkT}\right)^{1/2} I^{1/2} \tag{6.6}$$

and D is the dielectric constant of the medium and I the ionic strength. The quantity κ is the "reciprocal ion-atmosphere radius" from Debye–Hückel theory; it depends on the ionic strength of the medium. For large ionic strengths,

---

†It may be wondered why electrostatic units are used in these calculations. It is because calculation of force, energy, and other mechanical quantities in centimeter-gram-second units follows directly from the use of electrostatic units for electrical quantities. The reader is reminded that 1 V equals 1/300 esu of potential, and 1 C equals 3×10^9 esu of charge.

the atmosphere is shrunk tightly around the macroion; if I is small, the atmosphere is diffused and extended. The function $X(\kappa R)$ is called Henry's function; it varys between 1.0 and 1.5 as κR goes from zero to infinity (see Figure 6.2).

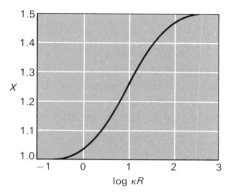

Figure 6.2 Henry's function. R is the macromolecular radius and κ the Debye–Hückel parameter.

[At this point it would be useful to work out Problem 3, so as to obtain a feeling for the quantities involved and to see the effect of the correction terms in Equation (6.5).] However, even Equation (6.5) does not provide an accurate description of the electrophoretic mobilities of real macroions. This is because no correction has been made for the fact that the ion atmosphere is distorted both by the field and by the motion of the particle through the medium [see Figure 6.1(*b*)]. More complex expressions have been derived, but the situation is still far from satisfactory. The net result of all of these complications is that electrophoresis, while providing an excellent method for separation of macromolecules, has proved less useful than methods such as sedimentation for obtaining information about macromolecular structure. It is exceedingly difficult, for example, even to obtain a reliable value for the macromolecular charge, z. Inaccurate as it may be, Equation (6.5) predicts one point of interest; at the pH at which a protein or other polyampholyte has zero charge (the isoelectric point) the mobility should be zero. At lower pH values a protein will be positively charged and move toward the negative electrode, and at higher pH transport will be in the opposite direction. Thus the measurement of isoelectric points can be made simply and unambiguously by means of electrophoresis. An example is shown in Figure 6.3. In this case, an attempt has been made to calculate the mobility of a protein at pH values different from the isoelectric point. Comparison of the result with the experimentally determined electrophoretic mobility demonstrates a considerable difference. Some of this may result from the binding of anions or cations in the buffer solution.

A number of techniques are available for electrophoresis; some are best for preparative separations, whereas others allow the precise control of conditions necessary for mobility measurement. The most accurate method in the latter respect is the *moving-boundary* method. The cell that is used is depicted in Figure 6.4; it is essentially a U-tube, with sliding joints that facilitate the forma-

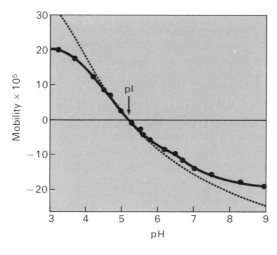

Figure 6.3 Mobility versus pH for the protein β-lactoglobulin. The points and the solid curve show experimental results, while the dashed curve shows the mobility predicted by Equation (6.5), using titration data for the charge. Taken from R. K. Cannan, A. H. Palmer, and A. C. Kibrick, *J. Biol. Chem.*, **142**, 803 (1942).

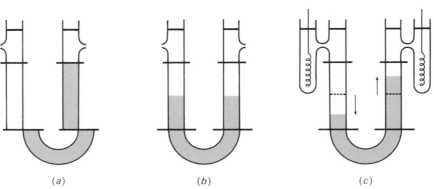

Figure 6.4 Formation of boundaries in a Tiselius electrophoresis apparatus. The U-tube is filled, as in (*a*), with solution (shaded) on one side and solvent (unshaded) on the other. In (*b*) the sliding partitions have opened and the two boundaries have formed. In (*c*) electrodes have been connected and electrophoresis has begun.

tion of sharp boundaries. The electrodes are contained in side arms, so that products of the electrolysis process do not fall into the boundary region. The boundaries between buffer and protein solution are formed as shown in the figure and then gently displaced into the straight limbs of the cell. These limbs are provided with flat windows in planes parallel to the paper. Through these windows the boundaries can be observed, using either a schlieren or interferometric optical system. The same apparatus, without the electrode compartments, can be used for diffusion studies (see Chapter 4).

To produce electrophoresis, a constant current is passed through the cell. Obviously, it would not be practical to directly compute the electrical field E

from the dimensions of the apparatus and the potential difference between the electrodes. Therefore, we make use of Ohm's law. Consider an element of volume, anywhere in the cell, of cross section A and thickness dx. If the potential difference across this slab is dV, then

$$dV = -i \, dR \qquad (6.7)$$

where i is the current and dR is the resistance of the slab of solution. (The minus sign corresponds to the convention that the current flows toward the lower potential.) However, we can write dR as dx/KA, where K is the specific conductance of the solution in this particular volume element. Furthermore, $-dV/dx$ is the potential gradient, E. Therefore,

$$E = \frac{i}{KA} \qquad (6.8)$$

Thus a knowledge of the current and the cross-sectional area of the cell, together with a separate measurement of the specific conductance, allows the calculation of E in either limb. The current is easily measured. It must be the same throughout the cell, and through the external circuit as well, or charge would accumulate at some point. Since observation of the motion of the boundary can be made with high precision, accurate mobility determinations can be made.

Although the moving-boundary method played a major role in the development of protein biochemistry, it is now mainly of historical interest and is rarely used. It is mentioned here because it is the only method available for the accurate determination of absolute mobility values. But we have seen that there is great difficulty in interpreting mobilities in terms of molecular parameters, and interest in their measurement has declined. Electrophoresis is one of the most widely used of all physical techniques at the present, but most applications are centered on analysis and separation, and the moving-boundary method is ill suited for either purpose.

6.2 Zonal Electrophoresis

The tremendous resurgence in the use of electrophoresis in recent years has been based almost entirely on what may be generally termed *zonal* techniques. In these methods a thin layer, or zone of the macromolecule solution, is electrophoresed through some kind of matrix. The matrix provides stability against convection, as does a sucrose gradient in zonal sedimentation. In addition, in many cases the matrix acts as a "molecular sieve" to aid in the separation of molecules on the basis of size.

The kind of supporting matrix used depends on the type of molecules to be separated, and the desired basis for separation: charge, molecular weight, or both. A list of some commonly used materials and their applications is given in Table 6.1. Almost all electrophoresis of biological macromolecules is presently carried out on either polyacrylamide or agarose gels. For this reason, the term *gel electrophoresis* is more commonly employed than "zonal electrophoresis."

TABLE 6.1 *Some media for zonal electrophoresis*

Medium	Conditions	Principal Uses
Paper	Filter paper moistened with buffer, placed between electrodes	Small molecules: amino acids and nucleotides
Starch gel	A bed of starch gel is laid horizontally between electrodes	Proteins, particularly isozyme separation; becoming obsolete
Polyacrylamide gel	Cast in tubes or slabs; cross-linked	Proteins and nucleic acids
Agarose gel	As polyacrylamide	Very large proteins, nucleic acids, nucleoproteins, etc.

These gels can be prepared in a wide range of concentrations, and in the case of polyacrylamide gels the density of cross-linking can also be readily controlled. For some purposes composite agarose–polyacrylamide gels are employed.

The principle of the technique is illustrated in the very simple application shown in Figure 6.5. The gel has been cast in a cylindrical tube; it contains the

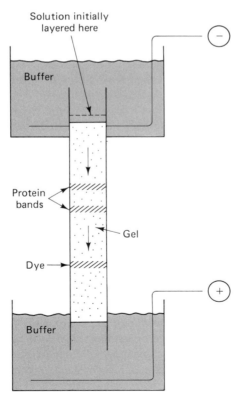

Figure 6.5 A schematic drawing of a very simple tube gel apparatus. Two negatively charged proteins have been separated during the electrophoresis.

buffer to be used. The macromolecular solution is applied in a thin layer at the top of the tube. When a voltage is applied, the zone of macromolecules migrates down the tube; if several components of different mobility are present, they will separate during electrophoresis. Usually, a dye of high mobility is added to the macromolecular solution; its migration serves to mark the progress of the experiment. The dye also serves as a convenient measure of mobility; the *relative mobility* U_{ri} of each component i is defined by

$$U_{ri} = \frac{U_i}{U_d} = \frac{d_i}{d_d} \tag{6.9}$$

where U_d is the dye mobility and d_i and d_d are the distances that component i and dye, respectively, have moved by the conclusion of the experiment.

It must be emphasized that the mobility exhibited by a macromolecule in gel electrophoresis is generally different from the mobility that would be found for the same substance, in the same buffer, in the moving-boundary electrophoresis experiment described above, that is, in "free electrophoresis." This is because there will always be a certain sieving effect. A macromolecule cannot move as easily through a molecular network as it can free in solution. In fact, a very simple relationship has been discovered between relative mobility and gel concentration, as shown in Figure 6.6. Such graphs are called *Ferguson plots*, after their devisor, H. A. Ferguson. They all obey the general equation

$$\log U_{ri} = \log U_{ri}^0 - k_i C \tag{6.10}$$

where C is the gel concentration and U_{ri}^0 is the relative mobility of component i when $C = 0$. Using Equation (6.9), we see that this should be related to the absolute mobility of i, as measured by the moving-boundary method, by

$$U_i = U_{ri}^0 U_d^0 \tag{6.11}$$

where U_d^0 is the absolute mobility of the reference dye. The constant k_i depends on the molecular weight; large molecules will have large values of k, whereas very small molecules will have small values, and hence will behave almost the same in a gel as they do in free electrophoresis. This is readily understandable, for to a small molecule the highly swollen gel matrix presents a resistance little different from that of the buffer solution.

With these few general concepts, we can explain much of what is observed in gel electrophoresis of proteins and nucleic acids. First, consider a miscellaneous group of proteins that are of various sizes and have various charges; their behavior will be as shown in Figure 6.6(*a*). Some are small but highly charged. These will have a large U_i^0 but a small k, as line 1. Others may be large, and also highly charged (line 2) or small, and have a small charge (line 3). Obviously, the mobilities of these proteins, relative to one another, will depend greatly on the gel concentration. In fact, the order in which they migrate on the gel can be changed by changing C. Interpretation is obviously difficult.

There are, however, some special cases in which interpretation is simplified. Consider, for example, the set of Ferguson plots shown in Figure 6.6(*b*). This is

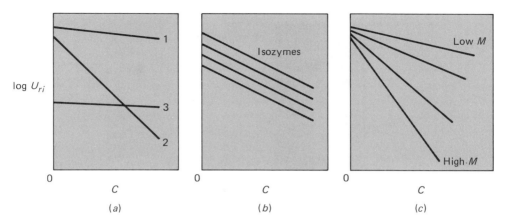

Figure 6.6 Ferguson plots for a number of commonly encountered situations.

what one would expect for a set of *isozymes*, protein molecules which have essentially the same mass and size, but differ in net charge. The Ferguson plots will be parallel lines.

An even more important case is that depicted in Figure 6.6(*c*), which clearly corresponds to a set of molecules which have essentially the same mobility in free electrophoresis but differ in size. How is this possible? One type of molecule that could approximate this kind of behavior is a long rod, with charge proportional to its length. From Table 4.1 we can show that the frictional coefficient of a thin rod of length L is given approximately by

$$f \simeq \frac{3\pi\eta L}{\ln (L/b)} \tag{6.12}$$

If the charge z is proportional to the length L, we have from Equation (6.3)

$$U \simeq \frac{KL}{f} \tag{6.13}$$

where K is a constant. Therefore,

$$U \simeq \frac{K}{3\pi\eta} \ln \left(\frac{L}{b}\right) \tag{6.14}$$

At large values of L/b, U will change very slowly with increasing L; in fact, as $L \rightarrow \infty$, $dU/dL \rightarrow 0$. Thus we may expect U to be essentially independent of molecular weight for a set of molecules of this kind. The same result applies to stiff coils such as DNA, where the charge is again proportional to the length. If the mobility in free electrophoresis is the same for all molecules, separation on the gel will depend *entirely* on the molecular sieving; therefore, the molecules will be separated on the basis of molecular size. This principle is widely used in the electrophoretic analysis of nucleic acids. Figure 6.7 illustrates the electrophoresis of a mixture of DNA fragments produced by restriction endo-

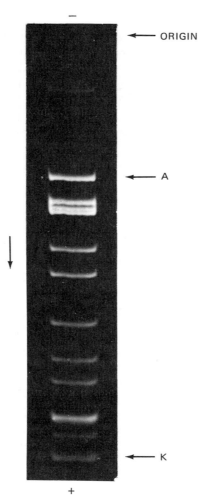

Figure 6.7 Gel electrophoresis of defined fragments of DNA produced by the action of the restriction endonuclease CfoI on the bacterial plasmid PBR 322.

nuclease digestion of a bacterial plasmid. Note the remarkably high resolution obtained. A graph of log M_i versus relative mobility is shown in Figure 6.8. For a given gel concentration, there will usually be some molecular weight range in which log M_i and U_{ri} are approximately linearly related. A series of fragments such as those shown in Figure 6.7, for which the molecular weights are exactly known from DNA sequencing, serves as an excellent "calibration set" for the estimation of the molecular weight of unknown DNA molecules. For a series of linear DNAs, mobility will depend only on M. But it must be remembered that it is molecular *size* that is of importance in the sieving effect, and changes in DNA conformation will modify the mobility. For example, a closed circular DNA will migrate at a different relative mobility than the linear form of the

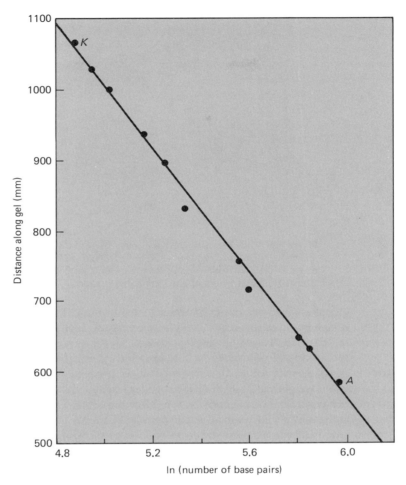

Figure 6.8 A graph of log M_i versus mobility for some of the fragments shown in Figure 6.7. The molecular weight, M, is expressed in base pairs; it is precisely known for each fragment. It will be noted that two fragments migrate anomolously. This is not unusual; there *are* specific sequence effects on mobilities.

same molecule. Remarkably, it is even possible to resolve circular DNA molecules that differ in the number of supercoil turns.

When a number of samples are to be compared, "tube gels" as depicted in Figure 6.5 are inconvenient. Most researchers now use a "slab gel" technique, with apparatus as shown in Figure 6.9. A number of notches are cast in the top of the gel slab by inserting a "comb" during polymerization of the gel. Each notch is loaded with a sample and provides a separate lane for electrophoresis.

The behavior shown in Figure 6.6(c) also serves as the basis for a very widely used technique for the determination of the molecular weights of polypeptide

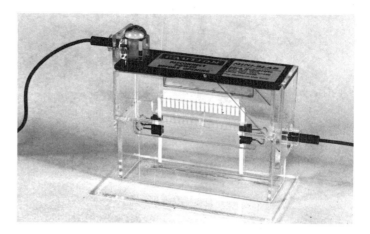

Figure 6.9 A commercially available slab gel electrophoresis apparatus. This apparatus is designed for small (8.3 cm by 10.3 cm) gel slabs, which can be 0.5 mm or 1 mm thick. Sixteen samples are run on each gel, with the comb shown here. Courtesy of Idea Scientific Co., Corvallis, Oregon.

chains—"SDS-gel electrophoresis." In this method, the protein to be studied is first denatured by heating with a detergent such as sodium dodecyl sulfate (SDS). Usually, a reducing agent such as β-mercaptoethanol is also added, to reduce any disulfide bonds. The protein is then electrophoresed in the presence of SDS. It is found that the separation proceeds on the basis of polypeptide chain weight, and is nearly independent of the charge on the polypeptide. This behavior can be explained on the following basis: The SDS apparently forms an elongated micelle, with the polypeptide chain as its core (see Figure 6.10). The length of this rodlike structure will be proportional to the polypeptide chain length and therefore to its molecular weight. The charged SDS molecules will be present in such excess in each micelle as largely to swamp the charge of the protein itself, and the amount of this SDS charge will also be proportional to the polypeptide chain length. Thus we once again have the necessary conditions for Equation (6.14), so molecules migrate according to their molecular weights. The size of an unknown polypeptide chain can be determined by calibration with chains of exactly known length. An example of the kind of resolution obtainable on SDS gels is shown in Figure 6.11.

While SDS-gel electrophoresis is a quick and inexpensive way to estimate chain weight, it should be used with caution, for a number of potential artifacts can lead to serious error. Some proteins, for example, bind more or less SDS than the average, which will make them run anomalously on the gel. If the protein itself has a very large positive or negative charge, this charge may not be negligible compared to the charge produced by the bound SDS. For example, the histones whose electrophoresis is shown in Figure 6.11 have a strong positive charge; if one attempted to measure their molecular weights from comparison

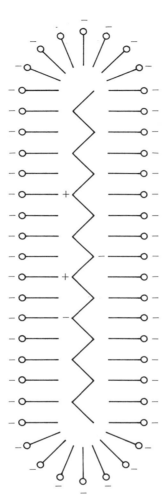

Figure 6.10 A schematic view of a postulated structure for SDS–protein micelles. The extended polypeptide chain lies at the center and is surrounded by a "bottle brush" of SDS molecules.

with "standard" proteins, the values found could be as much as 50 percent too high!

Gel electrophoresis techniques can also be employed on a preparative scale, to isolate particular nucleic acid fragments or polypeptides from complex mixtures. Since the technique is a zonal one, high purity can often be attained in a single step.

The high resolution seen in gels such as the one illustrated in Figure 6.11 is obtained by the use of a "stacking" gel and a discontinuous buffer system. To understand what these terms mean, consider the arrangement shown in Figure 6.12. On top of the "running gel" in the tube or slab is placed a "stacking gel" of lower concentration and thus higher porosity. The sample is layered on top of this. Note that quite different buffers are used in the upper reservoir, in the

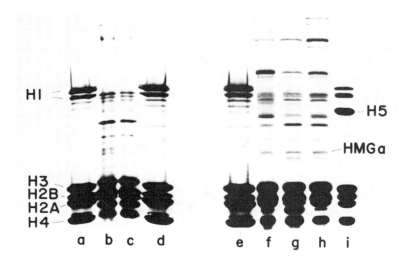

Figure 6.11 SDS-gel electrophoresis of acid-soluble nuclear proteins from a number of cell types. The major proteins present are the nucleosomal "core" histones, H3, H2B, H2A, and H4, plus the larger lysine-rich histones, H1 and H5. There are also a large number of minor nonhistone proteins. The gel has been somewhat overloaded with the histones in order to reveal these proteins. Lanes *a*, *d*, and *e* are from calf thymus; lanes *b*, *c*, *f*, *g*, and *h* are from various yeast strains. Lane *i* is from chicken erythrocytes. The samples were run on a 15 percent polyacrylamide slab gel, or apparatus similar to that shown in Figure 6.9. Migration is from the top toward the anode (bottom). The gel was stained with the dye Coomassie Blue and photographed. Courtesy of Dr. J. Davie.

stacking gel, and in the running gel. A critical feature for the operation of the system shown in the fact that the amino proton of glycine is only feebly dissociated below pH 9; the equilibrium

$$^+H_3NCH_2COO^- \quad \rightleftharpoons \quad H_2NCH_2COO^- + H^+ \tag{6.15}$$

lies well to the left, toward the electrically neutral zwitterion form. The apparent mobility of glycine under these conditions is very low. If a fraction α has dissociated, glycine molecules can be thought of as each spending a fraction of time α in the anionic form, and the observed mobility will be αU_{G-}, where U_{G-} is the mobility of anionic glycine.

The efficiency of the system in producing sharp bands depends on the fact that the mobility of glycine is made to be less than the mobility of chloride ion, by choosing the correct pH for the stacking gel. Proteins which are to be resolved into sharp bands should have mobilities intermediate between glycine and chloride. When the voltage is applied, and ions begin to move toward the anode, the Cl$^-$ ions will begin to move faster than the glycine. But any tendency for a "gap" to appear in the ion concentration will lead to a drop in conductivity in the boundary region. However, the current must be the same at every point in the gel column; therefore, the voltage gradient will increase across this boundary,

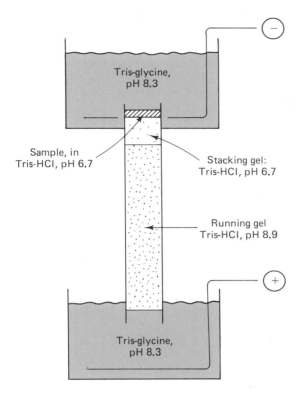

Figure 6.12 The principle of "stacking" in gel electrophoresis.

and the slower ions will be accelerated so as to keep up [see Equation (6.8)]. If only chloride and glycine were present, a sharp glycine–chloride boundary would move down the gel. The sharpness of this boundary is maintained, since as glycine falls back, it is accelerated by the increased potential gradient, and if it penetrates the chloride-containing solution, it is decelerated by the much weaker potential gradient existing in that medium of higher conductivity.

In the system as depicted, the protein molecules, which have intermediate mobilities, will be concentrated into a series of very thin "stacks" between the glycine and chloride. Each will be placed according to its relative mobility. Since the stacking gel is dilute, very little molecular sieving occurs here.

When the boundary passes from the stacking gel into the running gel, two important changes occur. First, the higher pH in the running gel favors the formation of more gly⁻ ion; therefore, the effective mobility of glycine increases. Second, the higher gel concentration in the running gel begins to impede the protein molecules. Consequently, glycine overpasses the proteins, and a glycine–chloride boundary moves ahead. The proteins are now electrophoresing in a uniform glycine buffer, and are being separated by the molecular sieving effect. Each protein stack entered the running gel as a very thin, concentrated layer, even though the proteins may have been present in a dilute solution when first applied to the gel.

The advantages of such a system are obvious. Rather sizable protein loads can be applied, and yet high resolution is possible because of the stacking phenomenon. We have examined one particular system for the sake of clarity, but there are many buffer systems which can be used for different requirements. The discontinuous-buffer stacking systems can be used with electrophoresis of proteins under nondenaturing conditions, or in the presence of SDS, for molecular size analysis.

In fact, the present range of gel electrophoresis techniques is truly enormous, with many special variants (use of urea, nonionic detergents, and so forth) for special applications. The reader is referred to the excellent manual by Hames and Rickwood (1981) for details. One major advance, however, should be noted particularly. It has become possible to improve resolution in many cases by the use of *two-dimensional gel* techniques. In such methods, as illustrated in Figure 6.13, a tube gel is run under one set of conditions. The tube is then placed at the

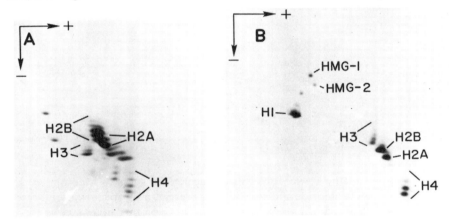

Figure 6.13 Two-dimensional gel electrophoresis of yeast histones (A) and calf thymus histones (B). The histones were first run on an SDS gel (horizontal direction) and then in an acetic acid–urea mixture (vertical direction). Note that the second direction allows the resolution of variants of several histones (for example, H4) which differ in charge because of amino acid modification. From J. Davie, *Anal. Biochem.*, **120**, 276–281 (1982). Reprinted by permission of the author and the publisher (Academic Press, Inc.).

top of a gel slab, and electrophoresis in the second dimension takes place under different conditions. The number of possible combinations is virtually limitless.

So far, we have said nothing about the important problem of locating the bands on a gel. There are a number of techniques in common use; they fall into three general categories:

1. The gel may be *stained*, using a dye that binds strongly to the macromolecules, and can be detected either by its absorption or, in some cases, fluorescence.

2. The gel may be *sliced*, and each slice assayed by whatever technique is

appropriate. For example, if the macromolecules are radioactive, the slices may be counted in a scintillation counter.

3. Alternatively, if radioactive materials are being used, *autoradiography* may be employed. The gel is placed in contact with a photographic film and left long enough to expose the film in those areas corresponding to the bands containing the radioisotope.

The use of autoradiography has permitted the development of some very elegant methods for the detection of specific substances in bands on slab gels. The Southern blotting technique, named for its inventor, E. Southern, will serve as an example. Suppose that one wishes to determine whether a particular DNA sequence is present in any of the bands that have been resolved on an agarose gel. The DNA is denatured in the gel with NaOH; the gel is then laid on a nitrocellulose sheet which is underlain with several sheets of filter paper. The single-stranded DNA molecules are then either squeezed (by pressure) or electrophoresed onto the nitrocellulose. Since lateral diffusion of the DNA is very slow, a replica of the gel pattern is created on the nitrocellulose. The sheet is then baked, to fix the single-stranded DNA molecules in place. It is then treated with a solution containing a radioactive "probe"—single-stranded DNA molecules of the sequence to be sought. Conditions are such that these can hybridize to any partner strands that may be affixed to the nitrocellulose. The sheet is then washed to remove unhybridized probe DNA, and subjected to autoradiography. Only those bands on the gel containing DNA fragments complementary to the probe DNA will be radioactive, and expose the film. This technique has proved enormously powerful in locating particular DNA sequences in enzymatic digests of DNA.

Two other techniques, facetiously termed "Northern" and "Western" blotting, have been developed to probe for particular RNA sequences or proteins in electrophorograms. The basic principle is similar to that used in Southern blots, and some features are given in Table 6.2. Details of these methods and references to the literature are given in Rickwood and Hames (1982) and Hames and Rickwood (1981).

TABLE 6.2 *Electrophoretic Transfer Techniques*

Method	Material Transferred	Transferred to:	Probed with:
Southern	Denatured DNA	Nitrocellulose	Radioactive DNA or RNA
Northern	RNA	Diazotized paper[a]	Radioactive DNA or RNA
Western	Proteins	Diazotized paper[a]	Radioactive antibody

[a]Paper to which has been coupled a diazo compound that will covalently bind to the RNA or protein.

6.3 Isoelectric Focusing

Gel electrophoresis is analogous to the techniques of band sedimentation. There also exists a technique that is the analog of the density gradient sedimentation equilibrium experiment. In this technique, called *isoelectric focusing*, components of different isoelectric points are brought to form stationary bands in a pH gradient. In the sedimentation equilibrium experiment, a particle moves to the point at which $(1 - \bar{v}\rho) = 0$ and then stops; in a pH gradient a charged particle will be moved by an electric field until it reaches a point where it is isoelectric. Its net charge will then be zero, and it will stop.

To carry out this kind of experiment, two criteria must be met:

1. Since bands are to be formed, which will be denser than the surrounding solvent, they must be stabilized against convection. This can be done either by carrying out the experiment in a sucrose gradient column, as shown in Figure 6.14, or by using a gel matrix.

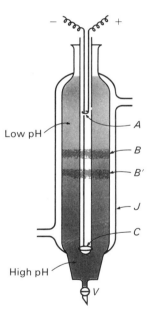

Low pH

High pH

Figure 6.14 A schematic drawing depicting the isoelectric focusing technique. The sucrose gradient is indicated by shading. The anode and cathode are platinum rings, denoted by *A* and *C*, respectively. Protein bands are shown formed at *B* and *B'*. The column may be emptied through the valve, *V*. It is cooled by the water jakcet, *J*. After H. Svensson, *Arch. Biochem. Biophys. Suppl.*, **1**, 132 (1962).

2. A very ingenious method has been devised to yield stable pH gradients. It is clear that after some electrolysis of a salt solution, the anode and cathode compartments of any electrophoretic apparatus will be acidic and basic, respectively. If a mixture of low-molecular-weight polyampholytes is placed in the intervening region, each will move to its isoelectric region and remain there, stabilizing the pH at a particular value at that point. The most acidic polyampholytes will be concentrated near the anode, and the most basic near the

cathode. A stable pH gradient will have been established. Now, any macro-molecule in the mixture will migrate to the region at which it is isoelectric. An equilibrium between electrophoresis and diffusion (resembling the equilibrium between sedimentation and diffusion in the sedimentation experiment) will result in concentration of each macromolecular species into a sharp band. Of course, this is not a true equilibrium, for electrolysis is continually going on, but the bands can be stable for rather long times. The technique can show remarkable resolution of proteins with closely spaced isoelectric points, as shown in Figure 6.15. The method can be utilized for either analytical or preparative purposes.

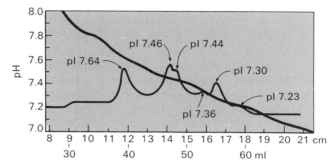

Figure 6.15 Separation of hemoglobin components by isoelectric focusing. The curve with peaks shows the absorption at 250 nm as a function of elutant volume and column position. The monotonic curve is the pH gradient. From A. Haglund, *Sci. Tools*, **14**, 17 (1967).

A particularly discriminating use of isoelectric focusing is in the O'Farell technique, which utilizes two-dimensional slab gels. The proteins are first separated by isoelectric focusing in a first dimension, and then subjected to SDS-gel electrophoresis in a second dimension. Thus the X and Y coordinates of a spot on the two-dimensional gel slab depend on the charge and molecular weight, respectively.

6.4 Orientation of Molecules in Electric Fields

So far, we have discussed the transport of molecules in electric fields. We now turn to another, and equally important effect—the orientation that is produced when the molecules have asymmetric charge distributions. A molecule in which charges $+q$ and $-q$ are separated by a distance d is shown in Figure 6.16. These

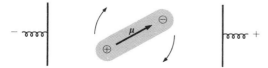

Figure 6.16 A dipole in an electric field. While there is no transport, since the net charge is zero, there is a torque tending to orient the molecule in the direction of the field.

charges might represent either ionized groups or simply the asymmetric charge distribution in a polar bond; in either case the molecule would be said to possess a *permanent dipole moment*. Alternatively, the displacement of charge might have been produced by an external electric field; in this case the dipole moment is *induced*. In either event, the dipole moment is a vector whose direction is given by the line connecting the centers of charge and whose magnitude† is defined as

$$\mu = qd \tag{6.16}$$

The vector nature of the dipole moment means that we can represent the dipolar character of even a very complex molecule as the single vector sum of all the moments of individual dipoles within the molecule. For example, a synthetic polypeptide in the α-helical conformation may have a very large permanent moment, because all of the small dipole moments of the amino acid residues along the chain add together, whereas the antiparallel double helix of DNA has no permanent moment but may have a very large induced moment. As Table 6.3 illustrates, the dipole moments of some macromolecules are indeed very large. These values should be contrasted with the moment for water, which is about 1.9D.

TABLE 6.3   *Dipole moments of some macromolecules*

Substance	*Solvent*	*M*	*μ (D)*
Myoglobin	Water	17,000	170
Egg albumin	Water	44,000	250
Horse carboxyhemoglobin	Water	67,000	480
Poly-γ-benzyl-L-glutamate	Ethylene dichloride	70,000	1,040
		102,000	1,570
		153,000	2,460

A molecule such as that depicted in Figure 6.16 will not be moved by an electric field since it has no net charge, but it will tend to be oriented. The analysis of this effect is not difficult. Consider the potential energy of the molecule in a field of strength E. The molecule in Figure 6.16 is depicted with its dipole moment at an angle θ with respect to the direction of the field; this places one charge center at a position x_1 and the other at a position x_2 in the field. The potential energy of a charge varies linearly with position in a homogeneous field, and the potential energy of the pair of charges can be considered the sum

†Dipole moments are usually expressed in debye units (D). A pair of charges (1×10^{-10}esu each) separated by 1 Å (10^{-8} cm) corresponds to a dipole moment of 1 D. Small polar molecules typically have dipole moments of the order of 1 D: asymmetric macromolecules may have moments of hundreds or thousands of debye units. Thus orientation experiments become feasible at low fields.

of the individual energies. Then we may write

$$
\begin{aligned}
V = V(x_1) + V(x_2) &= -(qx_2 - qx_1)E \\
&= -q(d \cos \theta)E \\
&= -\mu E \cos \theta
\end{aligned}
\tag{6.17}
$$

Thus the potential energy will be at a minimum when $\theta = 0$, that is, when the molecule is aligned with the field. The magnitude of this potential energy difference is directly proportional to the dipole moment, μ, and to the field strength, E.

We should not expect, of course, that the application of an electric field will produce total alignment of the dipolar molecules. The molecules will be constantly subjected to random interaction with solvent molecules, which will tend to produce random orientation. At equilibrium, a distribution of orientations should result; it is easy to calculate this from the familiar Boltzmann formula, Equation (1.10). If we call this distribution $f(\theta)$, then

$$
f(\theta) = K e^{\mu E \cos \theta / kT}
\tag{6.18}
$$

The normalization constant K can be calculated from the requirement that

$$
\int_0^{2\pi} d\phi \int_0^{\pi} f(\theta) \sin \theta \, d\theta = 1
\tag{6.19}
$$

Here ϕ is the rotation angle *about* the field axis, and the factor $\sin \theta$ appears because an increment of surface area of width $d\theta$ increases with θ (see Figure 6.17). Integration yields

$$
\frac{1}{K} = 4\pi \frac{kT}{\mu E} \sinh \left(\frac{\mu E}{kT} \right)
\tag{6.20}
$$

For small values of $\mu E / kT$, this reduces to $K \simeq \frac{1}{4}\pi$.

We can use this distribution function to calculate a number of important consequences of electric field orientation. For example, the absorption of polarized light by oriented molecules will be different if the plane of polarization is parallel or perpendicular to the applied electric field. We shall consider the simple case where the transition moment of the chromophore (see Chapter 8) is parallel to the permanent dipole moment. As shown in Chapter 8, the extinction coefficient observed for a chromophore that makes an angle θ with respect to the plane of polarization of the light is $\epsilon = \epsilon_0 \cos^2 \theta$, where ϵ_0 is the extinction coefficient when the transition moment is parallel to the polarization plane. Thus if we have a group of molecules distributed as in Equation (6.17), and illuminate them with light polarized parallel to the electric field, we should observe the extinction coefficient

$$
\epsilon_\parallel = \frac{\displaystyle\int_0^{\pi} \epsilon_0 \cos^2 \theta \, e^{\mu E \cos \theta / kT} \sin \theta \, d\theta}{\displaystyle\int_0^{\pi} e^{\mu E \cos \theta / kT} \sin \theta \, d\theta}
\tag{6.21}
$$

We assume here that $\mu E/kT$ is small, so that the factor K is simply a constant, which cancels. Evaluating these integrals gives, for small fields,

$$\epsilon_\| = \frac{\epsilon_0}{3}\left[1 + \frac{2}{15}\left(\frac{\mu E}{kT}\right)^2\right] \tag{6.22}$$

This shows that the application of the field enhances the absorption for light polarized parallel to the field, a result that might be intuitively expected for this case. In the absence of the field, the molecules are randomly oriented. As the field is increased, they tend to line up more and more parallel to the light polarization.

A similar calculation can be carried out for light polarized perpendicular to the electric field. In this case, the light absorption is reduced. The difference in extinction coefficients, divided by the extinction coefficient for unpolarized light, is defined as the *electric dichroism*

$$\frac{\Delta\epsilon}{\epsilon} = \frac{\epsilon_\| - \epsilon_\perp}{\epsilon} \simeq \frac{1}{10}\left(\frac{\mu E}{kT}\right)^2 \tag{6.23}$$

In the more general case, where the transition moment makes an angle α with the dipole moment, it can be shown that

$$\frac{\Delta\epsilon}{\epsilon} \simeq \frac{1}{10}\left(\frac{\mu E}{kT}\right)^2(3\cos^2\alpha - 1) \tag{6.24}$$

Thus electric dichroism measurements allow, in principle, the determination of the orientation of a particular chromophore in a macromolecule with respect to the dipole axis.

A closely related phenomenon is *electric birefringence*, or as it is sometimes called, the *Kerr effect*. Because of the close relationship between absorption and refraction, it is not surprising that oriented solutions of macromolecules will be birefringent, that is, display different refractive indices for light polarized parallel and perpendicular to the electric field. While we shall not go into the theory in detail here, it may be noted that the equations are, as would be expected, very similar to those for electric dichroism. For a rodlike molecule with dipole moment μ, the refractive index difference Δn is given by

$$\frac{\Delta n}{C} \simeq \frac{2\pi}{15m}\left(\frac{\mu E}{kT}\right)^2(\alpha_\| - \alpha_\perp) \tag{6.25}$$

Here $\alpha_\|$ and $\alpha_\perp$ are components of the optical polarizability parallel and perpendicular to the molecular axis, m is the molecular mass, n is the refractive index, and C is the concentration. Both electric dichroism and electric birefringence are exceedingly powerful techniques for studies of the anisotropy of rodlike macromolecules. The former technique is the more potentially fruitful, for it allows in principle the investigation of the relative orientation of various chromophoric groups on a given molecule, if these absorb at different wavelengths.

Further insight into the properties of macromolecules can be obtained by studying the *relaxation* of a partially oriented system to a state of random orientation when the field is suddenly turned off. Here we are concerned with a new type of nonequilibrium process, *rotational diffusion*. If we have a large collection of anisotropic molecules (let us assume them to be rods, for convenience), we may describe their distribution of orientations by a diagram such as Figure 6.17.

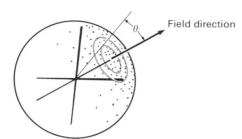

Field direction

Figure 6.17 Statistics of molecular orientation in an electrical field. Each point on the sphere surface represents the end of a rodlike molecule at a given instant. For random orientation the points would be uniformly distributed over the sphere surface. This is a very nonrandom distribution. The function $f(\theta)\,d\theta$ represents the fraction of molecules whose end points lie in the annular ring between θ and $\theta + d\theta$.

We are concerned only with the orientation with respect to a particular axis, z. Imagining all of the rods to be of the same length, and with their centers at the same point (0), we see that their ends will form a pattern of points on the surface of a sphere. The distribution $f(\theta, t)\,d\theta$ will then represent the fraction of points that lie in the annular ring between θ and $\theta + d\theta$, at a particular time t. The random rotational diffusion of orientations will then be represented by the random diffusion of points on the surface of a sphere. It is not hard to see that the process will in many ways resemble the process of translational diffusion. There will be an analog of Fick's first law for the flow of points across the circle defined by a particular θ:

$$J(\theta, t) = -\Theta\left(\frac{\partial f(\theta, t)}{\partial \theta}\right)_t \qquad (6.26)$$

The analog of Fick's second law is

$$\frac{\partial f(\theta, t)}{\partial t} = \Theta \frac{\partial^2 f(\theta, t)}{\partial \theta^2} \qquad (6.27)$$

In these equations, the quantity Θ is a *rotational diffusion coefficient*; like the translational diffusion coefficient, it will depend on the size and shape of the macromolecule.[†] For example,

$$\Theta = \frac{RT}{8\mathfrak{N}\pi\eta R^3} \qquad \text{(sphere of radius } R\text{)} \qquad (6.28)$$

$$= \frac{3RT}{16\mathfrak{N}\pi\eta a^3}\left(2\ln\frac{2a}{b} - 1\right) \qquad \text{(prolate ellipsoid, length } 2a, \text{ thickness, } 2b\text{)} \qquad (6.29)$$

The process of relaxation from partial orientation can now be described in the following way: Initially, after "long" (several milliseconds) application of the

[†]Note that the rotational diffusion coefficient depends on the *cube* rather than the first power of a dimension of the macromolecule. This makes it exceedingly sensitive to the lengths of asymmetric molecules.

field, the distribution of orientations will look somewhat like that shown in Figure 6.17; it will be described by Equation (6.17). When the field is suddenly removed, the rotatory diffusion process will relax the distribution to a random one, in which the end points in Figure 6.17 become uniformly distributed over the sphere. The process is very much like what would occur if we could suddenly turn off the centrifugal field on a system at sedimentation equilibrium. Without working out the result, we can guess that a property such as the birefringence or dichroism should relax exponentially with time toward zero. In fact, the equation for a property X that depends on the orientation is

$$X(t') = X_{eq}e^{-6\Theta t'} \tag{6.30}$$

where t' is the time after the field is turned off and X_{eq} the value at equilibrium in the field. Figure 6.18 shows a schematic drawing of such a process of orientation and subsequent relaxation. Thus, in addition to the other information

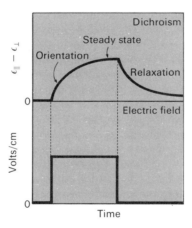

Figure 6.18 Orientation and relaxation in an electric dichroism experiment.

provided, electric birefringence or dichroism studies should yield rotational diffusion coefficients and hence data on molecular dimensions.

It must be understood that the discussion presented here applies to the simple case of a cylindrically symmetrical, rodlike macromolecule rotating about its short axis. Most real molecules will not possess such symmetry, and in the general case, *three* rotational diffusion coefficients, to describe rotation about the three axes, become necessary. Needless to say, all of the theories become correspondingly more complex.

PROBLEMS

1. The following data are for a moving-boundary electrophoresis experiment with gamma globulin: The descending boundary moved 1.40 cm in 136.5 min, through a solution with specific conductivity of 2.32×10^{-3} ohm^{-1}/cm. The cross-sectional area of the cell was 0.75 cm² and the current was kept at 20 mA. What is the electrophoretic mobility?

2. Consider two DNA molecules, one of which is 400 base pairs long, the other which is 800 base pairs long. Assuming that each is a rigid rod of diameter 25 Å, with a length of 3.4 Å per base pair, calculate the ratio of the electrophoretic mobilities of these two molecules, using the very approximate equation (6.3). Note that the charge will be proportional to the length.

3. At pH 7.0 in 0.1 M ionic strength buffer, at 4°C, normal adult hemoglobin has a mobility of about -0.2×10^{-5} cm²/sec-V, whereas sickle-cell hemoglobin has a mobility of about $+0.3 \times 10^{-5}$ cm²/sec-V. From these data, attempt to deduce the difference in charge between these molecules, using (a) Equation (6.4), and (b) Equation (6.5). An appropriate value for the radius can be estimated from Stokes' law, given $M = 6.8 \times 10^4$ g/mole, $\bar{v} = 0.75$ cm³/g, and $s_{20,w} = 4.3 \times 10^{-13}$ sec. Other useful data are $\eta = 10^{-2}$ poise at 20°C, 1.6×10^{-2} poise at 4°C, $D = 80$. It is now known that normal hemoglobin has a glutamate residue at position 6 of each β chain, whereas sickle-cell hemoglobin has a valine at this position. How does your result compare with this?

4. Digestion of chromatin by a nuclease such as micrococcal nuclease tends to cleave the DNA at the junctures between nucleosomes, to make a series of oligonucleosomes. To determine the DNA "repeat size," the DNA from such a digestion is electrophoresed on a polyacrylamide slab gel parallel to a lane of restriction fragments of a bacteriophage DNA of known sizes. The following data give distances migrated on such a gel:

Phage DNA Fragments		Fragments of Calf Nuclear DNA	
Size (base pairs)	*d (cm)*	*No.*	*d (cm)*
794	11.5	I	30.5
642	13.6	II	19.2
592	14.6	III	14.4
498	16.7	IV	11.5
322	22.5		
288	24.1		
263	25.4		
160	33.0		
145	34.2		
117	37.2		
94	40.0		

Deduce the repeat size of the calf chromatin.

5. A combination ultracentrifuge-electrophoresis apparatus could in principle be constructed by placing electrodes at the top and bottom of a centrifuge tube.

(a) Derive an equation for the velocity of motion of a particle of mass m and charge ze in such a combined field.

(b) If we imagine a sucrose gradient experiment in such an apparatus, it is clear that the sedimenting band could be stopped at any point by turning on an appropriate electrical field. For the normal adult hemoglobin described in

Problem 3, calculate the potential gradient (V/cm) needed to keep a band stationary at 6.5 cm from the center of rotation in a rotor turning at 60,000 rpm at 20°C.

(c)   Will the immobilized boundary be stable with time?

6.   Depending on the solvent used, a polypeptide may exist in different secondary structures. Consider the three possibilities: random coil, α helix, or antiparallel pleated sheet. Which of these would the data in Table 6.3 favor for poly-γ-benzyl-L-glutamate in ethylene dichloride? Explain your answer.

7.   Calculate the electric dichroism, $\Delta\epsilon/\epsilon$, in a field of 10,000 V/cm, for (a) a molecule with $\mu = 10$ D, and (b) $\mu = 1000$ D. Assume that the transition moment is perpendicular to the dipole axis in each case. Assume that $T = 300°$K. (c) At what value of α does $\Delta\epsilon/\epsilon$ change sign?

8.   An invertebrate hemoglobin is found to have a native molecular weight of about 70,000. Upon SDS-gel electrophoresis, the following results are found.

(1) After treatment with β-mercaptoethanol, the protein migrates as a doublet. The bands have traveled 10.0 and 10.6 cm.

(2) In the absence of β-mercaptoethanol treatment, only the 10.0-cm band of the doublet is seen, but there is a new band at 5.6 cm.

(3) A series of standard proteins migrates as follows:

Protein	M	d (cm)
Phosphorylase *b*	94,000	0.5
Bovine albumin	67,000	1.1
Ovalbumin	43,000	3.9
Carbonic anhydrase	30,000	6.6
Trypsin inhibitor	20,100	9.3
α-Lactalbumin	14,400	11.7

Describe the subunit structure of this protein.

9.   The electric birefringence of tobacco mosaic virus (TMV) decays to $1/e$ of its zero-time value in 5×10^{-4} sec at 20°C.

(a)   Calculate the rotational diffusion coefficient of TMV.

(b)   TMV is seen in the electron microscope as a rod about 180 Å in diameter by 3000 Å long. Predict the rotational diffusion coefficient, and compare with your result from part (a). The rotational diffusion coefficient of a cylinder is approximated by an equation exactly like (6.29), but with the factor -1 replaced by -1.6.

REFERENCES

Electrophoresis

Gaál, Ö., G. A. Medgyesi, and L. Vereczky: *Electrophoresis in the Separation of Biological Macromolecules*, John Wiley & Sons, Inc., New York, 1980.

Hames, B. D. and D. Rickwood (eds.): *Gel Electrophoresis of Proteins*, IRL Press Ltd., London, 1981. Excellent for practical details of a variety of techniques.

Rickwood, D. and B. D. Hames (eds.): *Gel Electrophoresis of Nucleic Acids*, IRL Press Ltd., London, 1982.

Weber, K. and M. Osborne: "Proteins and Dodecyl Sulfate: Molecular Weight Determination on Polyacrylamide Gels and Related Procedures," in *The Proteins*, 3rd ed. (H. Neurath and R. L. Hill, eds.), Academic Press, Inc., New York, 1975, Vol. I, pp. 179–223. A very fine review of this particular technique.

Electric Dichroism and Birefringence

Fredericq, F. and C. Houssier: *Electric Dichroism and Electric Birefringence*, Clarendon Press, Oxford, 1973.

Chapter Seven

VISCOSITY

The measurement of diffusion, sedimentation, and electrophoretic mobility provides considerable information about the size, shape, and gross conformation of dissolved macromolecules. However, it should be obvious that this information is far from complete and, in some cases, is beset with ambiguities in interpretation. To supplement these kinds of study, viscosity measurements are often used. To state their utility in an oversimplified way: The contribution of a macromolecular solute to the viscosity of a solution depends primarily on the effective volume occupied by the macromolecules. This is to be contrasted to the frictional coefficient, which depends on linear dimensions (a radius, for example; see Stokes' law). In this chapter we shall examine briefly the mechanics of viscous flow and show what studies of flowing solutions can tell about the macromolecular dimensions.

7.1 Viscous Flow

The resistance of fluids to flow is measured by their *viscosity*. To provide a precise definition of this quantity, it is necessary first to examine the mechanics of viscous flow. Consider the simple arrangement shown in Figure 7.1. A liquid is held between two infinite (very large) parallel plates. One of these is at rest; the other is pushed in the x direction with a constant velocity v. It is assumed

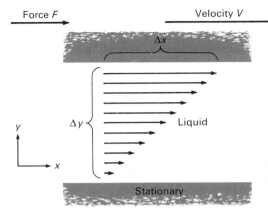

Force *F* Velocity *V*

Δ*x*

y

Δ*y* Liquid

x

Stationary

Figure 7.1 A schematic representation of the shearing of a Newtonian liquid between parallel plates. The arrows represent fluid velocities at different values of the coordinate *y*.

that the infinitesimal layer of liquid in contact with each plate "sticks" to it, so that the intervening layers of fluid must slide over one another in the manner shown. It is the degree of resistance to such sliding that is measured by the viscosity.

The kind of deformation that is experienced by the liquid sample in Figure 7.1 is referred to as shear. The *shear strain* at any moment is defined in this case as Δ*x*/Δ*y*. (Obviously, in this simple example the shear strain is everywhere uniform; were it not, we could define a shear strain at any point as *dx/dy*.) The *shear stress* is defined as the force pushing the top plate in the *x* direction, divided by the area (*A*) of the plate in contact with the sample. A liquid, when subjected to a constant shear stress, will shear at a constant *rate* as long as the stress is maintained.† For most liquids (Newtonian liquids) the relation between shear stress σ and rate of shear strain is very simple:

$$\sigma = \eta \frac{d}{dt}\left(\frac{dx}{dy}\right) \tag{7.1}$$

In this equation, η is the viscosity of the liquid; this expression may be taken as a definition of the viscosity. It may be written in a different way by changing the order of differentiation in Equation (7.1):

$$\sigma = \eta \frac{d}{dy}\left(\frac{dx}{dt}\right) = \eta \beta \tag{7.2}$$

Since *dx/dt* is the velocity of the liquid layer at height *y*, this formulation states that the viscosity is the ratio of the shearing stress to the gradient in fluid velocity

†The response to shear stress is very different for solids and liquids and provides a mechanical distinction between these states. An ideal solid, subjected to a constant shear stress, will simply assume an equilibrium value of the shear strain. Upon removal of the stress, the strain will vanish. Some substances exhibit combinations of the properties of both ideal solids and liquids. These are called *viscoelastic* materials; examples include most solid plastics, concentrated polymer solutions, and the like.

(β). These definitions specify the dimensions of η. Since stress is in dynes per square centimeter, viscosity has the dimensions of grams per centimeter-second; these units are called poise (P). The viscosity of water at room temperature is about 10^{-2} P, or 1 centipoise (cP).

The general conditions of flow described above correspond to *laminar* flow. They will usually be met in such experimental conditions as flow between concentric rotating cylinders or through a capillary tube as long as the rate of shear is not too great. At very high shear rates, turbulence will commence; turbulent flow is much more complex and will not be discussed here. The Newtonian viscosity law [Equation (7.1) or (7.2)] describes the flow of almost all low-molecular-weight liquids and some macromolecular solutions as well. The complications encountered when the shear rate is a more complex function of stress than given by Equation (7.1) will be discussed later.

7.2 Viscosities of Macromolecular Solutions

The interest of the biophysical chemist in viscosity measurements stems from the fact that the addition of some macromolecular solutes may greatly alter the viscosities of their solutions. While many theories of the viscosity of macromolecular solutions have been developed, all have their origin in the pioneering work of Albert Einstein. In 1906, Einstein developed the equation for the viscosity of a solution containing rigid, spherical solute particles. While the details of the derivation involve much sophisticated hydrodynamics, the general tenor of the argument can be presented easily. We begin by pointing out still another possible definition of viscosity: Viscosity is a measure of the rate of energy dissipation in flow. From Figure 7.1 it is clear that we may write the shear rate as $v/\Delta y$; if we multiply both sides of Equation (7.2) by this, we obtain

$$\frac{F}{A \, \Delta y} \frac{dx}{dt} = \eta \left(\frac{v}{\Delta y}\right)^2 \tag{7.3}$$

where the shear stress has been written out as a ratio of force to area. But $F \, dx$ is the energy expended in time dt in deforming the sample of fluid, and $A \, \Delta y$ is the volume of the fluid. Thus the left side of Equation (7.3) represents the rate of dissipation of energy per unit volume of fluid; this is seen to be proportional to the square of the shear rate, the factor of proportionality being η. That is,

$$\frac{dE}{dt} = \eta \left(\frac{v}{\Delta y}\right)^2 \tag{7.4}$$

Now consider a viscometer, like that in Figure 7.1, wherein we first measure the viscosity of a solvent by determining the shearing force necessary to produce a given rate of shear. We then introduce into the liquid a number of rigid particles; let us say that all together they occupy a fraction ϕ of the solution volume. We shall now find that a greater force is required to maintain the same shear rate:

$$\left(\frac{dE}{dt}\right)_s > \left(\frac{dE}{dt}\right)_0 \tag{7.5}$$

The rate of energy dissipation per unit volume is greater because in the solution there is a smaller volume of fluid in which the same overall deformation must be accomplished—the solute particles themselves cannot be deformed. This leads us to guess that the new rate of energy dissipation must be related to the rate in pure solvent by an equation of the form

$$\left(\frac{dE}{dt}\right)_s = \left(\frac{dE}{dt}\right)_0 (1 - v\phi)^{-1} \simeq \left(\frac{dE}{dt}\right)_0 (1 + v\phi) \tag{7.6}$$

which yields for the ratio of the viscosities

$$\frac{\eta_s}{\eta_0} = 1 + v\phi \tag{7.7}$$

Here v is a numerical factor, whose value we do not know. By a (vastly) more rigorous treatment of the problem, Einstein showed that for spherical particles $v = \frac{5}{2}$. For particles of other shapes, in dilute solutions, one finds an equation of the form of (7.7) with different values for v. Figure 7.2 shows the factor v as a function of axial ratio for prolate and oblate ellipsoids of revolution.

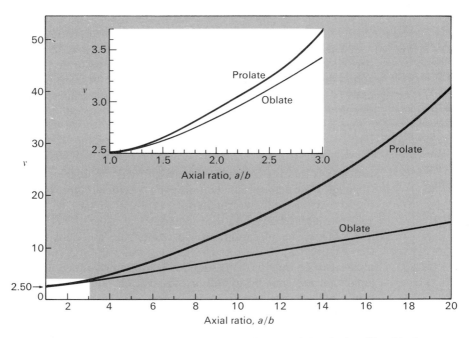

Figure 7.2 The viscosity factor v as a function of axial ratio for ellipsoids for revolution.

Equation (7.7) applies only in dilute solutions, for even its rigorous derivation neglects the possible effects of solute–solute interaction. These effects will give rise to quadratic and higher terms in the concentration, so that, in general, one should write

$$\eta_r = \frac{\eta_s}{\eta_0} = 1 + v\phi + \kappa\phi^2 + \cdots \tag{7.8}$$

Here we have introduced the term *relative viscosity* (η_r) for the ratio of solution to solvent viscosity. If we subtract unity from this, we obtain the *specific viscosity*, which is a measure of the fractional change in viscosity produced by adding the solute:

$$\eta_{sp} = \eta_r - 1 = \frac{\eta_s - \eta_0}{\eta_0} = v\phi + \kappa\phi^2 + \cdots \tag{7.9}$$

Generally, it is awkward to work in a volume fraction concentration scale, for one usually prepares solutions by weighing solute into a given volume of solution. We can rewrite Equation (7.9) by noting that $\phi = Cv$, where C is in grams per milliliter and v is the specific volume of the solute. Then

$$\eta_{sp} = vvC + \kappa v^2 C^2 + \cdots \tag{7.10}$$

or

$$\frac{\eta_{sp}}{C} = vv + \kappa v^2 C + \cdots \tag{7.11}$$

The limit of η_{sp}/C as $C \rightarrow 0$ is termed the *intrinsic viscosity*,† [η]; it will depend on the properties of the isolated macromolecules, since effects of interaction have been eliminated by extrapolation. According to Equations (7.9) and (7.11), one may determine [η] by measuring the relative viscosity at a number of solute concentrations and extrapolating η_{sp}/C to infinite dilution (see Figure 7.3). From Equation (7.11) we see that

$$[\eta] = vv \tag{7.12}$$

which says that the intrinsic viscosity of a macromolecule depends on its shape (through v) and its specific volume. The specific volume, however, is not that of the anhydrous polymeric material but must include, to some extent at least, solvent that is attached to the molecule. The value of 2.5 that v takes for spherical particles is a minimum; any deviation from sphericity leads to a larger value (see Figure 7.2). For proteins, the *anhydrous* specific volume is about 0.75 in most cases, so that we may estimate the minimum value for the intrinsic viscosity (spherical proteins, no hydration) to be about 2 ml/g. As can be seen from the results collected in Table 7.1, a number of proteins approach this value, testifying

---

†Note that relative viscosity, specific viscosity, and intrinsic viscosity do not have the dimensions of viscosity at all. The first two are dimensionless ratios, while [η] has units of milliliters per gram.

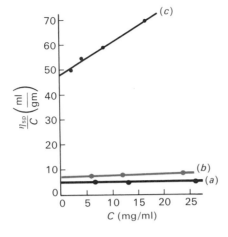

Figure 7.3 Determination of the intrinsic viscosity of crab hemocyanin in various solution environments. (*a*) Data for the native hemocyanin ($M \simeq 1 \times 10^6$) at neutral pH. (*b*) Subunits of the hemocyanin ($M \simeq 7 \times 10^4$) at pH 10.6, (*c*) Unfolded subunits in $6M$ guanidine hydrocloride, pH 10.6.

TABLE 7.1   *Viscosity behavior of macromolecules*

Substance	M	$[\eta]$ (ml/g)	$\beta \times 10^{-6}$	Θ (sec^{-1})
Globular particles				
Ribonuclease	13,683	3.4	2.01	
Serum albumin	67,500	3.7	2.04	6.4×10^5
30S *E. coli* ribosomes	900,000	8.1		
Bushy stunt virus	10,700,000	3.4		
Rods				
Fibrinogen	330,000	27	2.15	4.0×10^4
Myosin	440,000	217		
Tobacco mosaic virus	39,000,000	36.7	2.61	3.4×10^2
Poly-γ-benzyl-L-glutamate[a]	340,000	720		
Coils				
Poly-γ-benzyl-L-glutamate[b]	340,000	184		
Serum albumin[c]	67,500	52		
Myosin subunits[c]	197,000	93		

[a]Measured under conditions such that the molecule is an α helix.
[b]Measured under conditions such that the molecule is a random coil.
[c]Measured in 6 M guanidine hydrochloride; the molecule is a random coil.

to their somewhat compact shape. On the other hand, such elongated, rodlike molecules as DNA and myosin exhibit much higher values.

The principal difficulty in interpreting intrinsic viscosity data for globular proteins lies in the fact that both hydration and asymmetry are contributing to the same quantity. If one assumes that water of hydration has about the same density as ordinary water, then a hydration of δ grams of water per gram of dry protein will yield $v \simeq \bar{v}(1 + \delta \bar{v}_0/\bar{v})$, where $\bar{v}$ is the partial specific volume of anhydrous macromolecular solute, and $\bar{v}_0$ is the partial specific volume of water. Therefore,

$$[\eta] = v(\bar{v} + \delta \bar{v}_0) \tag{7.13}$$

If δ is known, the axial ratio can be calculated from v (although one still does not know whether the molecule is a prolate or oblate ellipsoid or even if it remotely resembles an ellipsoid). On the other hand, given v from separate knowledge about shape, one may obtain δ. Neither of these procedures has turned out to be especially fruitful, although sometimes the combination of a number of kinds of measurement (intrinsic viscosity, frictional coefficient, and rotatory diffusion coefficient, for example) can provide enough redundancy of information to allow intelligent guesses about both shape and hydration.

A more direct attack on this problem is provided by the formulation of H. A. Scheraga and L. Mandelkern (1953). These authors have combined sedimentation coefficient (s), intrinsic viscosity, and molecular weight to yield the equation

$$\frac{\mathfrak{N}s[\eta]^{1/3}\eta_0}{(100)^{1/3}M^{2/3}(1 - \bar{v}\rho)} = \beta\left(\frac{a}{b}\right) \tag{7.14}$$

where $(1 - \bar{v}\rho)$ is the familiar buoyancy term in sedimentation theory, $\mathfrak{N}$ is Avogadro's number, η_0 is solvent viscosity, and $\beta(a/b)$ is a function of the shape of the particle. A graph of β versus axial ratio (a/b) is shown in Figure 7.4. It will

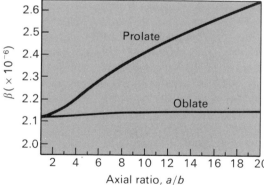

Figure 7.4 The factor β as a function of axial ratio for ellipsoids of revolution. The limiting value of β for an infinitely thin oblate ellipsoid is 2.15×10^6. Thus any molecule with a larger β must be prolate.

be noted that this combination possesses the peculiar advantage that β behaves rather differently for prolate and oblate particles. Thus, given supplementary information, one should be able to make a decision about particle shape. That all is not straightforward, however, is indicated by the values of β listed in Table 7.1. A number of apparently reliable results fall below the minimum value of 2.12×10^6 predicted by the theory. The Scheraga–Mandelkern equation has been used as a way for estimating the molecular weights of substances such as very high-molecular-weight nucleic acids, for which the more conventional techniques are not very satisfactory. The relative insensitivity of β to shape makes a good guess for β possible for this calculation.

The intrinsic viscosities of solutions of randomly coiled macromolecules are often found to be vastly larger than those for compact structures (see Figure 7.3). The reason for this is not hard to see, in a qualitative way. Such molecules occupy a very large effective volume in solution, and intrinsic viscosity is primarily a measure of particle volume. We can arrive rather easily at an expression for $[\eta]$ by assuming that the effective radius of the polymer coil will be of the same order of magnitude as the radius of gyration, R_G. If Equation (7.12) is rewritten on a volume-per-molecule rather than a volume-per-gram basis, it becomes

$$[\eta] = v\mathfrak{N}\frac{v_m}{M} \qquad (7.15)$$

where v_m is the molecular volume and M the molecular weight. Now if we assume that $v_m = K''R_G^3$, where K'' is a constant, we obtain an expression of the form

$$[\eta] = K'\frac{R_G^3}{M} \qquad (7.16)$$

with K' another constant.

For a series of samples of a given polymer, the radius of gyration will be a function of the chain length and hence of the molecular weight. If the polymer is dissolved in a θ solvent (see Chapter 2), the radius of gyration will be proportional to the square root of the molecular weight [Tanford (1961)]:

$$[\eta] = KM^{1/2} \qquad (7.17)$$

Thus in this case the intrinsic viscosity will depend on the square root of the molecular weight. If the solvent is "better" than a θ solvent, the radius of gyration will increase more rapidly with M. Then an equation of the form

$$[\eta] = KM^a \qquad (7.18)$$

will obtain, where a is some number greater than $\frac{1}{2}$. These predictions are supported by the data summarized in Table 7.2. Note that whenever a θ solvent is used, $a = 0.5$, as predicted. Equations of this type are exceedingly useful to the polymer chemist, for they allow an easy determination of molecular weights from intrinsic viscosity measurements. However, the parameters K and a must always be determined empirically for each solvent–solute system, using for calibration samples of known molecular weight. This dependence of the intrinsic viscosity on M, it should be emphasized, is peculiar to random-coil polymers and results from the way in which their effective radius depends on M. For compact particles there is no direct dependence of $[\eta]$ on molecular weight. Equation (7.15) empha-

TABLE 7.2 *Intrinsic viscosity-molecular weight relationships for polymers:* $[\eta] = KM^a$

Polymer	Solvent	$T (°C)$	$K \times 10^2$	a
Polystyrene	Benzene	25	0.95	0.74
	Cyclohexane	34[a]	8.1	0.50
Polyisobutylene	Benzene	24[a]	8.3	0.50
	Cyclohexane	30	2.6	0.70
Amylose	0.33 N KCl	25[a]	11.3	0.50
Poly-γ-benzyl-L-	Dichloroacetic acid	25	0.28	0.87[b]
glutamate	Dimethylformamide	25	1.4×10^{-5}	1.75[c]
Various proteins	6 M guanidine hydrochloride	25	0.716[d]	0.66
	+0.1 M β-mercaptoethanol			

[a]θ solvents.
[b]Coil form.
[c]Helical form.
[d]The equation is given in the form $[\eta] = Kn^a$, where n is the number of amino acid residues. See Tanford et al. (1967).

sizes that $[\eta]$ depends on the ratio of molecular volume to molecular weight. As shown in Table 7.1, some virus particles have intrinsic viscosities as small as the smallest proteins, while smaller asymmetric molecules may have large values of $[\eta]$.

When a globular protein molecule is denatured, its conformation is altered to something rather like that of a random coil. There is an accompanying increase in the effective volume in solution, so that $[\eta]$ generally increases. This can be a convenient way to study denaturation processes, as illustrated in Figure 7.5. Tanford et al. (1967) have investigated a number of proteins in guanidine–HCl solutions. These results attest to the unfolded state of the polypeptide chain (see Table 7.2).

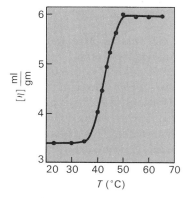

Figure 7.5 The change in intrinsic viscosity at pH 2.8 of the protein ribonuclease upon thermal denaturation. The value at low temperature is typical for a compact globular protein; the increase reflects the partial uncoiling of the molecule at high temperatures. The process is reversible, for upon cooling the value of $[\eta]$ again decreases.

7.3 *Orientation in Viscous Flow*

If an elongated macromolecule is placed in a liquid subjected to shear, there will be a torque exerted that tends to orient the molecule so that its long axis is parallel to the direction of flow (see Figure 7.6). In many respects, the situation is similar to that described for a dipole in an electric field (Chapter 6). The random Brownian rotational motion of the molecule will oppose such orientation, and the effect of a constant shear rate will be to produce an equilibrium distribution of orientation states. Note, however, that this is not a true equilibrium but rather a steady state, since energy is constantly being dissipated in the medium.

The theory is a good deal more complicated than that for orientation in an electric field. In the first place, as can be seen from Figure 7.6, it is necessary to

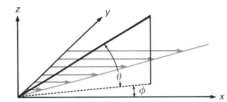

Figure 7.6 The coordinate system used to describe the orientation of an elongated molecule in a shear gradient.

describe the orientation with respect to two angles, ϕ and θ. Most results that have been obtained are given in terms of series solutions, valid only at low rates of shear. For the fraction of molecules with orientation angles between θ and $\theta + d\theta$ and ϕ and $\phi + d\phi$, the approximate expression has been obtained:

$$\rho(\phi, \theta) = K\left(1 + \frac{\beta}{\Theta} \frac{A \sin 2\phi \sin^2 \theta}{4} + \cdots\right) \qquad (7.19)$$

This equation applies specifically to ellipsoids of revolution with semiaxes a and b. The shear gradient [see Equation (7.2)] is denoted by β, Θ is the rotational diffusion constant, and A depends on the axial ratio of the molecule: $A = (a^2 - b^2)/(a^2 + b^2)$. The fact that powers of A are involved in the series expansion means that the result is restricted not only to small shear rates but also to relatively small axial ratios. Calculations have been carried to higher terms, but application of the theory rapidly becomes quite cumbersome.

Orientation by shear manifests itself in optical properties of the medium, much as does orientation by electric fields. Shear orientation possesses the advantage that it is not restricted to macromolecules with large dipole moments or large polarizability but suffers in application from the complexity of the theory and the fact that relaxation experiments, which have proved so fruitful in electric birefringence studies, are much more difficult. However, much useful information has been gained from the shear analogs of electric dichroism and birefringence. Results from shear dichroism studies of F-actin are shown in

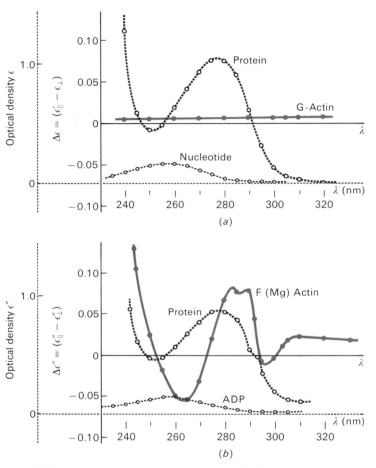

Figure 7.7 Flow dichroism of (*a*) globular actin and (*b*) fibrous actin. In each case the absorption spectrum is shown by the open circles with broken lines, and the dichroic spectrum by the closed circles and full lines. Polymerization of the non-dichroic, globular actin monomers leads to fibers that exhibit strong dichroism. The results show that the aromatic residues of the protein are oriented preferentially with transition moments parallel to the polymer axis, while the nucleotide transition moments are perpendicular to the axis. From A. Wada (1964). Reprinted with permission of author and publisher (John Wiley & Sons, Inc. Copyright 1964.)

Figure 7.7. This fibrous protein exists as a complex with adenosine diphosphate (ADP). When the solution is oriented by shear, the directions observed for the polarization of the absorption by the ADP and the aromatic side chains of the protein differ, with the result that the dichroism changes in sign in the vicinity of 275 nm.

Birefringence in shear is usually not studied by direct measurement of the refractive index difference for light polarized in different directions but by determination of a quantity called the *extinction angle*. To understand this, examine the experimental arrangement depicted in Figure 7.8. Here are shown concentric cylinders with their axes perpendicular to the paper and the solution placed in the annular gap between them. If one cylinder is rotated and the other is held at rest, a shear gradient will be produced in the liquid. Now suppose that we look down through the space between the cylinders and place a polarizer and an analyzer above and below the apparatus, respectively. If these are rotated so that their directions of polarization are perpendicular, no light will be transmitted through the system. If the cylinder is now made to turn, producing shear and orienting the molecules, the birefringence so produced will allow light to once more be transmitted—but not everywhere. At four positions around the circle the average orientation of the macromolecules will be parallel to the direction of polarization of either the analyzer or the polarizer. At these points the solution acts simply as an extension of the polarizer or analyzer, so light does not get through. Thus we shall observe a crosslike pattern, as shown in Figure 7.8. The angle χ that this pattern makes with the polarizer directions will depend on the orientation of the particles. It is now obvious why this angle is called the extinction angle. If we have a theory for the distribution of molecular orientations, this angle can be predicted. Using the distribution given by Equation (7.19), χ becomes

$$\chi = 45° - \frac{1}{12} \frac{\beta}{\Theta} \frac{180}{\pi} + \cdots \qquad (7.20)$$

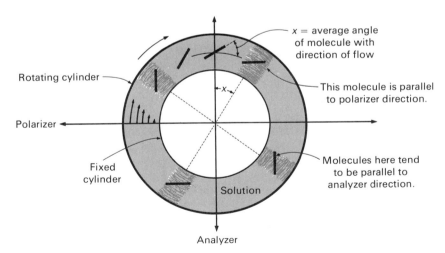

Figure 7.8 This is a schematic view, looking down on an apparatus for flow birefringence experiments. The orientations of "typical" molecules are shown by the rods in the solution. Note that extinction occurs only when the molecular axes are parallel to either the polarizer or analyzer directions.

Thus observation of the flow birefringence in this way allows the evaluation of the rotational diffusion constant from the limiting slope of χ versus β at low β. As pointed out in Chapter 6, the rotational diffusion constant is a useful quantity for it depends primarily on the length of a rodlike molecule. Some results obtained from orientation in shear are summarized in Table 7.1.

At low rates of shear, χ should approach 45°, although its observation becomes difficult, as the pattern fades with the decrease in orientation. The limit of χ at infinite shear, where all of the molecules should be oriented parallel to the direction of flow, should be 0°, but it is difficult to reach very high shear rates without turbulence, which destroys the orientation.

A final effect of the orientation of macromolecules in shear is a decrease in the viscosity itself. We have so far assumed that Newton's law [Equation (7.1)] is of general validity, but it is easy to see why solutions of elongated macromolecules might behave as *non-Newtonian* liquids. As the solute molecules are aligned by the gradient, the resistance that they contribute to the sliding of layers of liquid past one another decreases. Theories have been developed to predict this effect, again in terms of the distribution of orientations. A very clear example of the effect of particle shape on non-Newtonian behavior is shown in Figure 7.9. The polypeptide molecules in the helical conformation orient easily,

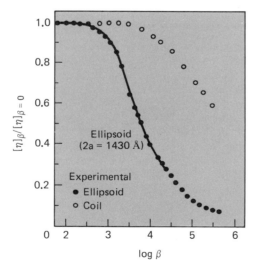

Figure 7.9 The non-Newtonian behavior of poly-γ-benzyl-L-glutamate in *m* cresol (lower curve) and dichloroacetic acid (upper curve). In the former case it is a rigid helix, in the latter a random coil. After J. T. Yang, *J. Am. Chem. Soc.*, **80**, 1783 (1958). Copyright by American Chemical Society.

as one would expect them to. The same sample when placed in a solvent in which the random-coil structure is favored demonstrates a much smaller degree of orientation.

The non-Newtonian nature of many polymeric solutions can be an experimental nuisance in the determination of intrinsic viscosity. If one employs a simple capillary viscometer (see Section 7.4) with a moderately high rate of shear for the study of, say, DNA solutions, a value for the intrinsic viscosity may be obtained that is much smaller than the value that would be found at low shear rate. In such studies it has become routine to extrapolate $[\eta]$ to zero rate of shear.†

7.4 Measurement of Viscosity

As can be seen from Equation (7.9), the determination of the intrinsic viscosity of a macromolecule requires the measurement of the relative viscosities of dilute solutions. Since the relative viscosity of a protein solution containing 1 mg/ml may be as low as 1.003, it is obvious that very high precision in the viscosity measurement is required if the value of $[\eta]$ is to have any reliability at all. Capillary viscometers are well suited to this purpose, for while they do not easily yield absolute viscosity values, they can provide a very accurate comparison of the viscosity of a solution and solvent.

The simplest and most common type of capillary viscometer is the Ostwald viscometer, shown in Figure 7.10(a). One simply measures the time required for a given volume of liquid (as measured by marks on the upper bulb) to flow through the capillary. The volume rate of flow of a liquid through a capillary of radius a and length l, when driven by a pressure P, is given by Poiseuille's law:

$$\frac{dV}{dt} = \frac{\pi a^4 P}{8\eta l} \tag{7.21}$$

Here dV/dt is in cubic centimeters per second and η is the viscosity. The liquid is driven, of course, by the hydrostatic pressure head. $P = hg\rho$ (g is the gravitational constant and ρ the liquid density). The height h varies slightly between limits h_1 and h_2 as the bulb empties. Integrating Equation (7.21) yields an expression for the flow time:

$$t = \frac{\eta}{\rho}\left(\frac{8l}{\pi g a^4}\int_{h_1}^{h_2}\frac{dV}{h}\right) \tag{7.22}$$

It is best to regard all of the factors in parentheses on the right of Equation (7.22) (including the integral) as instrument constants. Then, for the measurement of a relative viscosity, we may use the simple expression

$$\frac{\eta_s}{\eta_0} = \frac{t_s}{t_0}\frac{\rho_s}{\rho_0} \tag{7.23}$$

where the subscript zeros refer to solvent properties. According to this equation, the calculation of relative viscosity follows directly from the ratio of flow times

†In the case of very high-molecular-weight DNA, a further complication would appear at high shear. It is found that the viscosity *irreversibly* decreases as successive experiments are performed. This is because high shear gradients can physically break very long molecules. For this reason, high-molecular-weight DNA solutions should not be run through a narrow pipette or syringe.

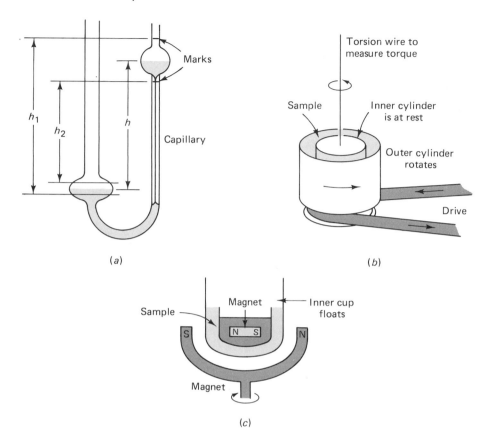

(a)   (b)

(c)

if the density ratio is known. The measurements can be made very precisely, but it should be pointed out that good temperature control is needed, for the viscosity is highly temperature dependent. (A difference of 1°C makes about a 2 percent difference in the viscosity of water; this is a difference comparable to many specific viscosities.)

The capillary viscometer has the disadvantage that the rate of shear cannot be very easily changed or even precisely specified. The pattern of flow through a capillary tube is such that the shear gradient is greatest at the walls and zero in the center. This means that only an average shear gradient can be specified and that it is determined by the apparatus constants and decreases during the flow (as the driving head falls). For measurements of non-Newtonian liquids, these complications are annoying. While capillary viscometers have been built that have several bulbs at different heights to provide several average shear rates, or are driven by an externally controlled air pressure, the capillary viscometer is rarely used for precise investigation of shear-rate dependence. Instead, concentric cylinder viscometers of the type shown in Figure 7.10(*b*) are most commonly employed, devices very similar to those already described for flow birefringence studies. One of the cylinders is driven at a constant speed, and the torque exerted on the other is measured. If the ratio of the gap between the cylinders to cylinder diameter is very small, the conditions approximate those of flow between infinite plates (Figure 7.1), and a nearly uniform rate of shear is obtained.

A simple and accurate version of this viscometer is shown in Figure 7.10(*c*). In this device the inner cylinder is a self-centering float, containing a small magnet. It is driven by an external revolving solenoid, and the viscosity of the liquid in the annular space causes the float to rotate with a constant velocity, determined by the dimensions of the apparatus and viscosity of the solution. This apparatus (called the Zimm viscometer) has become widely used in studies of DNA.

When *very* large DNA molecules are studied in an apparatus like that shown in Figure 7.10(*c*) a new phenomenon is observed. If the driving magnet is suddenly stopped, the inner cylinder does not simply cease to rotate; it actually reverses and turns in the opposite direction. Such DNA solutions are *viscoelastic*, and strain energy has been stored in deforming the immense DNA molecules. As these molecules relax to their equilibrium conformations, they drag the rotor backward. The relaxation can be approximately described by a very simple equation. If $\theta(t)$ is the angle through which the rotor has been turned at any time t during the relaxation, and $\theta(\infty)$ is the limiting value at very long time,

$$\ln \left[\theta(t) - \theta(\infty)\right] \simeq -\frac{t}{\tau} \tag{7.24}$$

where τ is termed the *relaxation time*. At 25°C, the relaxation time is empirically found to be related to the molecular weight by

$$\tau = \left(\frac{M}{1.45 \times 10^8}\right)^{5/3} \tag{7.25}$$

It is evident from Equation (7.25) that the method should be most useful for very large DNA molecules. First, note that even a sample of $M \simeq 10^8$ will have a relaxation time of less than 1 sec. Furthermore, since τ depends on the $\frac{5}{3}$ power of M, the relaxation time rapidly increases for DNA molecules larger than this.

The method has proved most useful in estimating the molecular weight of truly giant molecules, such as the DNA from an entire chromosome. Some such data are given in Table 7.3. No other physical technique is capable of studying macromolecules with molecular weights in the neighborhood of 10^{10}. From such experiments it has been concluded that each chromosome probably contains a single immense DNA molecule.

TABLE 7.3 *Viscoelastic relaxation times for some large DNA molecules*

Molecule	M	τ (sec)
Phage T7 DNA	2.5×10^7	0.05
Phage T2 DNA	1.1×10^8	0.57
E. coli DNA	2.5×10^9	70
Tetrahymena pyriformis DNA	1.5×10^{10}	2,100
Drosophila melanogaster	$\{\ 4.3 \times 10^{10}$	6,000
Chromosome DNA	$\{\ 5.9 \times 10^{10}$	10,700

PROBLEMS

1. The following data describe the determination of the intrinsic viscosity of bovine serum albumin (BSA) in a solvent containing 33.3 per cent dioxane and 0.03 M KCl at pH 2.0, $T = 25°C$. The protein concentration (C) is in grams per 100 ml; ρ/ρ_0 is the ratio of solution density to that of the solvent.

C	ρ/ρ_0	Flow Time Readings (sec)		
0.000	1.000	398.1	398.2	398.2
0.417	1.0011	439.3	439.1	439.1
0.685	1.0019	467.6	467.6	467.5
0.844	1.0023	485.8	485.5	485.7

(*a*) Calculate $[\eta]$.

(*b*) What would you conclude about BSA under these conditions? Compare with the results in Table 7.1.

2. According to P. Flory, the intrinsic viscosity of a random-coil polymer should be proportional to R_G^3/M, and the frictional coefficient should be proportional to R_G. From this, find a product of powers of s and $[\eta]$ that should provide a measure of M.

3. A spherical, unhydrated protein molecule has a radius of 25 Å and $\bar{v} = 0.740$. It dimerizes to form a dimer that is a prolate ellipsoid 100 Å long and of the same *volume* as two of the monomers. Calculate the percent change in s and $[\eta]$ upon dimerization and decide which method would be best for following the process.

4. Prove that

$$\lim_{C \to 0} \frac{1}{C} \ln \eta_r = [\eta]$$

and hence that $[\eta]$ may be determined by plotting $(1/C) \ln \eta_r$ versus C.

5. The hemoglobin of the clam, *Astarte castanea*, has very large polypeptide chains, of molecular weight 335,000. Their sedimentation coefficient at 20°C is 11.3S, and the intrinsic viscosity is found to be 6.87 cm³/g. The partial specific volume can be taken to be 0.743 cm³/g.

(a) Calculate the Scheraga–Mandelkern parameter β. Is the molecule best described by an oblate or prolate ellipsoid? Estimate the axial ratio.

(b) Given this value of the axial ratio, estimate the hydration of the protein.

(c) As a cross-check, predict the value of f/f_0 predicted for a prolate ellipsoid, and compare with the experimentally observed value of 1.49.

6. What value of the shear gradient would be required to produce an extinction angle of 40° with tobacco mosaic virus? (See Problem 9, Chapter 6.) If the shear were produced in a concentric cylinder apparatus, in which the inner (stationary) cylinder had a radius of 5 cm and the outer (moving) cylinder a radius of 5.5 cm, approximately what radial velocity (revolutions per minute) would be required?

7. The largest human chromosome has been estimated to contain a DNA molecule of molecular weight about 1.6×10^{11}. If this could be studied intact by viscoelastic relaxation, estimate the relaxation time that would be observed.

REFERENCES

Viscometry

Scheraga, H. A. and L. Mandelkern: *J. Am. Chem. Soc.*, **75**, 179–184 (1953).

Tanford, C.: *Physical Chemistry of Macromolecules*, John Wiley & Sons, Inc., New York, 1961, Chap. 6.

Tanford, C., K. Kawahara, and S. Lapanje: *J. Am. Chem. Soc.*, **89**, 729–736 (1967).

Yang, J. T.: "The Viscosity of Macromolecules in Relation to Molecular Conformation," *Adv. Protein Chem.*, **16**, 323–400 (1961). Very complete, although somewhat dated with respect to techniques.

Flow Dichroism

Wada, A.: "Dichroic Spectra of Biopolymers Oriented by Flow," *Appl. Spectrosc. Rev.*, **6**, 1–30 (1972).

Viscoelastic Relaxation

Kavenoff, R., L. C. Klotz, and B. H. Zimm: *Cold Spring Harbor Symp. Quant. Biol.*, **38**, 1–8 (1974). An application of this method to the study of whole-chromosome DNA molecules; references to earlier work.

Chapter Eight

ABSORPTION AND EMISSION

OF RADIATION

In the remaining chapters, we shall be dealing with interactions between radiation and molecules. As the reader has surely learned, electromagnetic radiation cannot be uniquely described on either a wave or particle basis; it must be considered to have some of the attributes of both. In the various ways in which matter and radiation interact, there are some that are most easily described by emphasizing the wavelike aspects (scattering, for example) and other processes, such as absorption, that compel us to consider the existence of photons. In what is to follow, we shall unashamedly use whichever point of view is most convenient at the moment, but the reader is cautioned that a most rigorous treatment of *all* the phenomena would be from a quantum-mechanical basis. We do not intend here to go deeply into either classic electromagnetic theory or quantum mechanics. Instead, we shall be concerned with the applications of scattering and absorption phenomena to the investigation of biological macromolecules.

8.1 Absorption Spectra: General Principles

In classical physics the absorption of radiation will occur when the frequency of the electromagnetic wave is close to or equal to a *natural vibration frequency* of the molecule. There are occasions when this analogy between mechanical phenomena and absorption will be useful, but in general it will not get us very far.

This point of view cannot explain the existence of *sharp* spectral lines, nor does it relate the *amount* of energy absorbed by an atom or molecule to the frequency of the radiation.

The discussion of absorption processes, then, requires that we adopt the point of view of quantum mechanics. Basically, there are two ways in which this differs from the classical picture:

1. The requirement that the energy available for absorption be related to the frequency (v) of the radiation by Planck's law:

$$E = hv \qquad (8.1)$$

where h is Planck's constant (6.67×10^{-27} erg-sec). There is no equivalent to this in classical physics; it imposes, as it were, a particulate aspect on radiation.

2. Recognition of the fact that atoms and molecules can exist at only certain *energy levels*. These turn out to be natural consequences of a wave-mechanical picture of matter.

We are not concerned here with the details of how the allowed energy levels of a system are deduced. However, the results for two simple examples that throw light on much more complex systems will be presented.

In classical physics a particle can have any energy. However, in quantum mechanics we find that constraining a particle automatically restricts its energy to only certain allowed levels. A specific, if somewhat abstract, example is shown in Figure 8.1. The particle (an electron, for example) is held in a one-

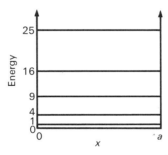

Figure 8.1 The simplest quantum-mechanical system—a particle in a one-dimensional box. The potential energy is zero within the box ($0 \leq x \leq a$) and infinite outside ($x > a$ or $x < 0$). The horizontal lines show the allowed energy levels.

dimensional "box" between $x = 0$ and $x = a$. Wave mechanics then requires that the particle can have only energies

$$E_n = \frac{n^2 h^2}{8ma^2} \qquad n = 1, 2, 3, 4 \ldots \qquad (8.2)$$

where m is the mass of the particle and n a quantum number. Absorption of radiation could then occur only for frequencies corresponding [by Equation (8.1)] to transitions between these levels. Note that the spacing of the allowed levels is greater the smaller the box. We shall make use of this general principle later.

A more complex system (but still simple by biochemical standards) is a diatomic molecule. Here we must make distinctions between the different kinds of energy the system can have. In the first place the energy of the molecule will depend on the electronic state—specifically, the set of orbitals that the electrons in the molecule occupy. For a given state, the energy will depend on the distance between the nuclei, in a fashion similar to that shown by curves I or II in Figure 8.2. For each electronic state, our molecule will have a set of allowed levels of

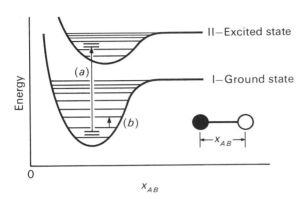

Figure 8.2 Energy levels for a simple diatomic molecule. The potential energy curve is a function of the internuclear distance (X_{AB}). Two electronic states are shown (I and II). They differ in their potential energy and in the position of the minimum, the equilibrium internuclear distance. For each electronic state there are different possible levels of vibrational energy (long lines) and rotational energy (short lines, shown only for two vibrational levels). A possible electronic transition is shown by (*a*) and a vibrational transition in the electronic ground state by (*b*).

vibrational energy (the horizontal lines in Figure 8.2). Finally, the *rotational* energy is also quantized; these states would correspond to sets of even more closely spaced lines clustered above each vibrational level.†

These simple systems allow us to demonstrate the most important features of absorption spectroscopy. Transitions may occur between electronic, vibrational, or rotational levels. The energy change, and hence the frequency of the radiation involved, will generally decrease in the same order. Figure 8.3 schematically describes the electromagnetic spectrum and indicates the general ranges in which each type of transition is to be found. Note that the wavelength (λ) or frequency (ν) can be expressed in a number of ways. For wavelength, one usually uses either angstrom units (Å), micrometers or microns (μm), or nanometers (nm). Frequency is given in sec^{-1} (since $\nu = c/\lambda$), but sometimes the *wave number* $\nu' = 1/\lambda$ is given instead, usually in cm^{-1}.

Spectroscopy, to the biochemist or biophysicist, generally means the observation of absorption spectra. Except in the cases of fluorescence and phosphorescence, *emission* of electromagnetic radiation is appreciable only at temperatures too high to preserve the integrity of sensitive biological molecules. Most studies are made on solutions, and these are quite often dilute solutions. In such circumstances the simple Beer–Lambert law will govern the absorption processes:

$$I = I_0 e^{-\epsilon' lc} = I_0 \times 10^{-\epsilon lc} \tag{8.3}$$

where I_0 is the intensity of the incident radiation and I the intensity of the radia-

†For a more detailed analysis of such quantum-mechanical problems, the student is referred to any of the better physical chemistry texts.

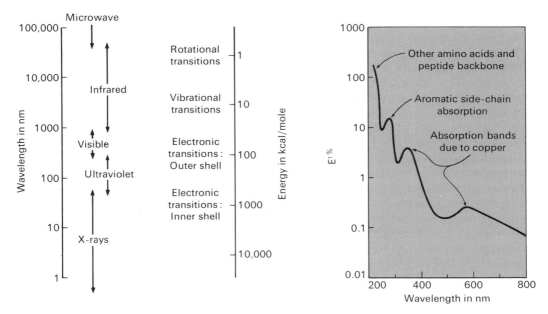

Figure 8.3 (*a*) A sketch of the electromagnetic spectrum. (*b*) The visible and near-UV spectrum of the copper protein, hemocyanin. Bands in the visible and near-UV arise from the copper moeity; those near 280 nm from aromatic protein side chains.

tion transmitted through a cell of thickness *l* cm, containing a solution of concentration *c* moles per liter. The quantity ϵ is the *extinction coefficient*, with the units liter mole^{-1}cm^{-1}. Data are frequently reported in percent transmission ($I/I_0 \times 100$) or in absorbance [$A = \log(I_0/I)$]. The latter is particularly convenient, for from Equation (8.3),

$$A = \epsilon l c \qquad (8.4)$$

Sometimes the extinction coefficient is given in other units; for example,

$$A = E^{\%} l C \qquad (8.5)$$

where the concentration *C* is in grams per 100 ml of solution. This is useful when the molecular weight of the solute is unknown or uncertain.

The measurement of absorption spectra, then, requires that we be able to produce monochromatic light of known wavelength and measure the decrease in intensity that occurs when this light passes through a known thickness of solution. Today such intensity measurements are almost always made electronically, using either photocells or photomultiplier tubes. A sketch of a modern spectrophotometer is shown in Figure 8.4. The light source produces a wide range of radiation; the wavelength required is selected by a prism or grating

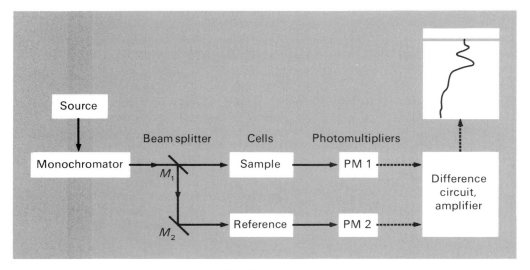

Figure 8.4 A simplified diagram of a recording spectrophotometer.

monochromator. The beam is split, and the two portions traverse, respectively, the sample cell and a reference cell, containing only solvent. The intensities are measured by photomultiplier tubes, and the difference between the logarithms of sample and reference intensity is recorded. In many such instruments the wavelength scale is scanned automatically, with A recorded as a function of λ on a synchronized recorder.

8.2 *Infrared Spectra*

In Section 8.1 we sketched some general principles of spectroscopy. These ideas apply even to the largest molecules of interest to the biochemist, albeit with certain restrictions and complications.

In the first place, pure rotational spectra are rarely observed for biochemically interesting substances. Most of these materials are studied in solution; in this dense environment, pure rotational states cannot exist as they do in the gas.†

Turning to the shorter-wavelength region of the near infrared, we find an embarrassing wealth of detail in the absorption spectra of even the simpler molecules of biological interest. The reason for this is clear; here we are observing vibrational transitions and should be considering the possible normal modes of vibration of the molecules. A nonlinear molecule with n atoms will have $3n - 6$ fundamental modes of vibration. Even for a simple substance such

†While pure rotational spectra are not observed for molecules in solution, there exists low-frequency absorption that apparently corresponds to *librational* motions, which are rotational oscillations about equilibrium positions. This area of spectroscopy has been little explored.

as an amino acid, this yields a large number; for a macromolecule such as protein, a complete analysis of the infrared spectrum is quite beyond present imagining. Nevertheless, the fact that certain groups display vibrational transitions at characteristic frequencies makes some unscrambling possible. For example, the —C=O group has a fundamental stretching frequency of about 1700 cm^{-1} in many molecules, and a fundamental stretching mode of the —N—H group is almost always observed in the neighborhood of 3400 cm^{-1}. A list of such frequencies is given in Figure 8.5. An example of the kind of

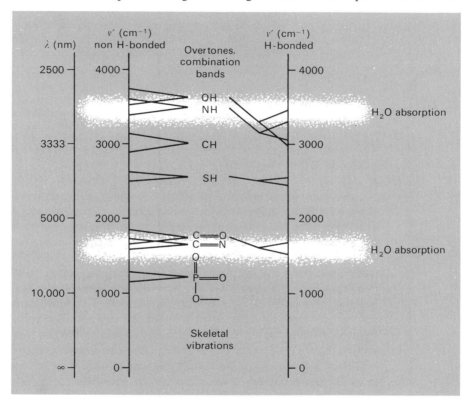

Figure 8.5 Some typical infrared absorption bands found in biological molecules. Approximate shifts on H-bonding are shown, and the regions of water absorption are shown within the white area.

information that can be obtained by the study of such identifiable infrared bands is shown in Figure 8.6. The polypeptide is in the α-helical configuration so that the —C=O groups are hydrogen-bonded. This results in a decrease of the stretching frequency from the value of about 1700 cm^{-1} for free —C=O groups to a value (1652 cm^{-1}) close to that observed in crystalline amides, where hydrogen bonding can be demonstrated by X-ray diffraction.

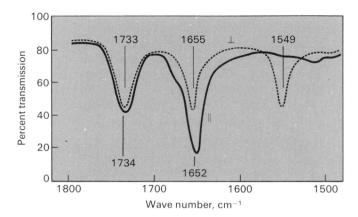

Figure 8.6 Infrared spectrum of a film of poly-γ-benzyl-L-glutamate in the α-helical form. A solid film of the polymer has been prepared and oriented by stretching. The dashed line is for radiation polarized perpendicular to the molecular orientation and the solid line for radiation polarized parallel to the helix axes. Adapted from M. Tsuboi, *J. Polymer Sci.*, **59**, 139 (1962). Copyright © 1962, John Wiley and Sons, Inc.

A similar study is shown in Figure 8.7, which demonstrates the base pairing of nucleoside derivatives in nonaqueous solution. The observation of a maximum absorbance for the hydrogen-bond bands at 50 mole percent *U* shows that a 1 : 1 complex is formed.

A great impediment to the use of infrared analysis of biological materials is the very strong absorption by water in the regions around 3400 and 1600 cm⁻¹ (see Figure 8.5). This prohibits, for example, an otherwise easy study of hydrogen bonding of macromolecules in aqueous solution. To some extent, this difficulty can be overcome by using dried films of the substance or by working in D_2O solutions, where the solvent absorption is shifted to the less interesting regions around 1200 and 2500 cm⁻¹. However, one can never be sure that the subtle details of macromolecular structure that are so important in biochemical processes are not modified by such changes in molecular environment. An alternative is to use Raman spectra (see Chapter 9).

A good deal of useful information has resulted from studies of oriented films of elongated macromolecules using polarized infrared radiation. Under such circumstances, dichroism will be observed; the absorption will depend on the relative orientation of the absorbing chromophoric group and the direction of polarization of the light. We can express this quantitatively by considering, as in Figure 8.8, that the *transition dipole moment* (very approximately, the charge displacement that accompanies the absorption) makes an angle θ with respect to the direction of light polarization. If the amplitude of the electromagnetic wave is represented by a vector **E** (see Chapter 9), the projection of this vector in the direction of the transition moment will be proportional to

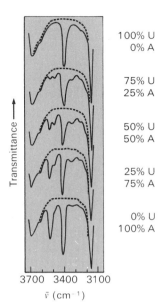

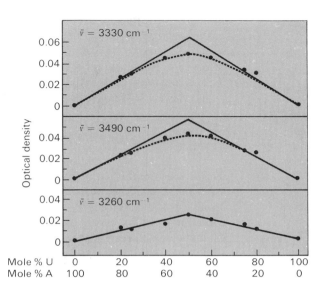

Figure 8.7 Infrared detection of the hydrogen-bonded association of 1-cyclohexyl uracil and 9-ethyladenine in deuterochloroform solutions. *Left*, spectra (dashed lines are background) for different mixtures. *Right*, intensities of bands as a function of composition. The bands at 3,330, 3,490, and 3,260 cm^{-1} are indicative of hydrogen-bonding. After R. M. Hamlin, R. C. Lord, and A. Rich, *Science*, **148**, 1734 (1965). Copyright © 1965, by the American Association for the Advancement of Science.

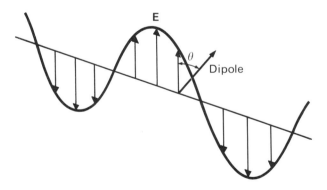

Figure 8.8 The basis of dichroism. The angle between the direction of polarization and the direction of the transition dipole determines the probability of absorption.

cos θ. The energy absorbed will be proportional to the intensity, or the *square* of this component of the amplitude. Thus the extinction coefficient, ϵ_θ, for light polarized at an angle θ to the transition moment will be given by

$$\frac{\epsilon_\theta}{\epsilon_\parallel} = \cos^2 \theta \qquad (8.6)$$

where $\epsilon_\parallel$ is the extinction coefficient for light polarized parallel to the transition moment. Note that when $\theta = 90°$, $\epsilon_\theta = 0$. This effect of orientation can be seen in the spectrum of poly-γ-benzyl-L-glutamate shown in Figure 8.6. The —C=O stretching band (~ 1650 cm^{-1}) is polarized strongly along the axis of elongation of the film, showing that the corresponding bonds must lie parallel to this axis. This is consistent with the idea that the molecules are in the form of α helices in which these bonds are supposed to lie parallel to the long axis of the molecules.

8.3 Spectra in the Visible and Ultraviolet Regions

Transitions in the visible region of the spectrum are relatively low-energy *electronic* transitions. Roughly speaking, there are two main kinds of biological structures that have energy levels with spacings in this range: (1) compounds containing certain metal ions (particularly transition metals), and (2) large aromatic structures and conjugated double-bond systems. The former include some metalloproteins and the latter such conjugated structures as vitamin A. Heme proteins owe most of their intense absorption in the visible region to the conjugated heme group rather than the metal ion. It will be noted that electronic absorption bands are usually quite broad. This is because they actually consist of a large number of closely spaced subbands, each corresponding to a different change in vibrational energy accompanying the change in electronic state (see Figure 8.2). In macromolecules such as proteins additional broadening results from the fact that individual chromophores of a given type may exist in a number of different environments.

In the near-ultraviolet region of the spectrum (200 to 400 nm), many biochemicals exhibit strong and useful absorption bands. In keeping with the rough "particle in a box" rule, it is not surprising that these are often associated with *small* conjugated ring systems, both aromatic and heterocyclic, whereas the large conjugated systems absorb in the visible region. Thus among the amino acids, tyrosine, phenylalanine, and tryptophan absorb most intensely in this region (Table 8.1). Most of the absorption of proteins in the 280-nm region results from the presence of these residues. Similarly, all of the purine and pyrimidine bases absorb strongly in the near ultraviolet. Most of these absorption bands appear to correspond to transitions of π electrons in the ring to antibonding π orbitals (the so-called π–π^* transitions). Evidence for this is seen in Figure 8.9, where spectra of crystals of 1-methylthymine are shown.

Farther in the ultraviolet, nearly *everything* absorbs. Suppose we consider the absorption spectrum of a hypothetical protein. The 280-nm absorption is due, of course, to the aromatic side chains, but below about 230 nm we begin to encounter not only transitions in other amino acid side chains but also those involving electron displacements in the peptide backbone itself. There appears to be a particularly intense absorption of the latter type at about 190 nm and a weaker one at about 220 nm. These transitions will be discussed in greater

TABLE 8.1 *Near-ultraviolet absorption bands*
of some amino acids and nucleotides[a]

Substance	p.H	λ_{max} (nm)	ϵ (liters/cm·mole)
Phenylalanine	6	257	200
Tyrosine	6	275	1,300
Tryptophan	6	280	5,000
Adenosine-5′-phosphate	7	259	15,000
Cytidine-5′-phosphate	7	271	9,000
Uridine-5′-phosphate	7	262	10,000
Guanosine-5′-phosphate	7	252	14,000

[a]In each case the longest wavelength band is given. Each compound has other (and generally stronger) bands in the vicinity of 200 nm.

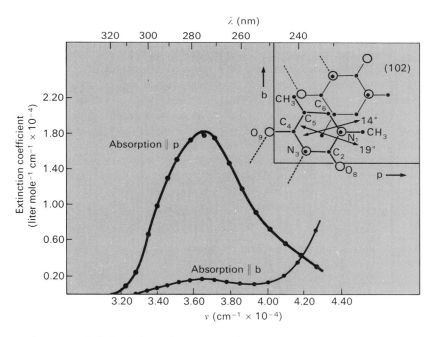

Figure 8.9 Polarized absorption spectra of crystals of 1-methylthymine. The absorption is from π–π^* transitions; probable directions of transition moments are shown. After R. F. Stewart and N. Davidson, *J. Chem. Phys.*, **39**, 255 (1963).

detail in Chapter 10. Below about 170 nm lies *terra incognita*; work in this region is rendered exceedingly difficult by the absorption by oxygen of the air, in addition to that by water and almost all other solvents. Experiments seem possible only with thoroughly dried materials in evacuated spectrographs—hence the term *vacuum ultraviolet*.

8.4 *Macromolecular Structure and Absorption*

So far, we have made no distinction between the spectra of macromolecules and those of their monomers. As a first approximation this is justified, especially in the case of random-coil polymeric structures. However, when a definite secondary structure is established in a macromolecule, certain changes are commonly observed in the spectrum. In many cases these have been used as diagnostic tools to detect and follow configurational changes.

Consider, for example, the following effect: When a protein is hydrolyzed to its constituent amino acids, the spectrum of the products is somewhat different from that of the intact protein. Similar changes are observed when the protein is *denatured*, that is, when its definite secondary and tertiary folding are disturbed. These observations suggest that the environment of the absorbing residues must be different in the folded protein molecule than when the amino acids are in free contact with the solvent. While the theory is still somewhat incomplete, the effect can be measured with considerable accuracy by *difference spectrophotometry*. In this technique the optical density difference between the substance in two states is measured directly (using, for example, a double-beam spectrophotometer). Thus the change can be used to follow such processes as denaturation.

With nucleic acids even more dramatic effects are observed. Figure 8.10

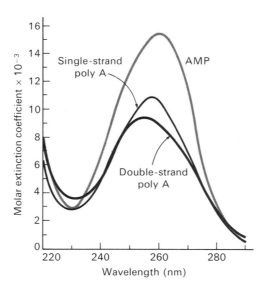

Figure 8.10 Hypochromism of polyriboad-enylic acid. Note that the form of the polymer spectrum differs little from that of the monomer but that the intensity of absorption is considerably reduced.

contrasts polyriboadenylic acid with an equal concentration of its monomer, adenosine monophosphate, while Figure 8.11 shows the change produced in the spectrum of *E. coli* DNA by heating to 90°C with accompanying denaturation. Let us consider the simpler case of the synthetic polyriboadenylic acid. Two

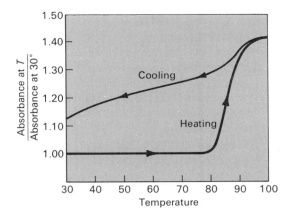

Figure 8.11 The "melting" of DNA as followed by absorption at 260 nm. Note the sharpness of the transition in cycle 1. This is typical of a cooperative process (see Chapter 3). Also note that recovery of the ordered structure is not complete on rapid cooling.

effects are observed; the polymer shows a *lower* absorption intensity than the monomer (hypochromism), and in the polymer the maximum is shifted to a wavelength almost 3 nm lower.

Both these effects have been explained on the basis of the helical configuration of the nucleotide polymers. In such a regular, closely packed structure the chromophores cannot be considered to be independent of one another. The electronic displacement that occurs when one base absorbs a quantum of energy is felt in the neighboring bases as a modification of the electrical field. If the interaction is strong, it becomes impossible to speak any longer of an absorption "localized" in a particular residue. Rather than the single, N-fold degenerate absorption frequency that a polymer of N independent groups would exhibit, an *exciton band*, with N closely spaced but distinct levels, is now observed. The total absorption intensity will not necessarily be distributed equally among the various levels; in this way a shift in the maximum wavelength of absorption may be produced.†

A simple example will illustrate the principle. Suppose we have a dimer of a substance that in the monomeric form has an absorption at the frequency v_0. If in the dimer there is interaction of energy U between the chromophores, the single monomer band will be split into two bands at frequencies

$$v_1 = v_0 + \frac{U}{h} \qquad v_2 = v_0 - \frac{U}{h} \tag{8.7}$$

(see Figure 8.12). The relative absorption intensities will depend on the geometric relation between the transition moments in the two chromophores. In a real system the vibrational broadening of the bands would probably prevent resolution, and we might observe only a shift in the frequency of maximum absorption.

†Of course, this is not the only way in which a change in absorption frequency can arise. Simply putting a chromophone into a different environment (inside a helix, for example) may change its electronic properties.

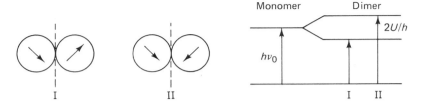

Figure 8.12 Splitting of a transition upon dimer formation. The two modes of excitation in the dimer lead to different excited-state energies.

The hypochromism of polymeric structures also arises from intramolecular interactions. Due to interactions with the solvent and surrounding polymer, some of the transitions of the monomer groups may be more favored and others less. Thus the hypochromism of the 260-nm band in polyadenylic acid or DNA, and its change with conformation, is not surprising. A very general rule of spectroscopy holds that mere configurational modifications that do not change the fundamental electronic structure should not change the *total* absorption, integrated over all bands. If this is to hold for the nucleic acids in question, we must postulate an equal *hyperchromism* somewhere else, presumably in the inaccessible low-wavelength region.

The hypochromism in the ultraviolet bands of the nucleic acids has become an exceedingly useful tool for the observation of configurational change since spectroscopic absorption measurements are rapid and precise.

8.5 Fluorescence and Phosphorescence

In discussing absorption we have been concerned entirely with the excitation of a molecule from a ground state to a higher energy level and have given no consideration to the subsequent fate of the excited molecule. In many cases the sequel is not very interesting; the energy is transferred, as heat, to the surroundings. However, in some cases reradiation occurs and this will usually be at a different frequency than that of the exciting light. This process is called *fluorescence*.

Figure 8.13(*a*) shows, in a crude fashion, how a substance can exhibit fluorescence. If the molecule is in the lowest vibrational level of the electronic ground state, its internuclear distance will most likely be about that shown by point *a* in the figure. This is dictated by the form of the vibrational wave function, which will give a maximum probability for this internuclear distance. If an electronic excitation occurs, it will happen so quickly that no appreciable motion of the massive and sluggish nuclei will have taken place (the Franck–Condon principle). Thus we should expect the transition to be described by a nearly vertical line (*ab* in the figure), and the most likely transitions will be those that lead to vibrational states that have maximum probability of internuclear distance close to the initial distance. In the example shown, the form of the potential

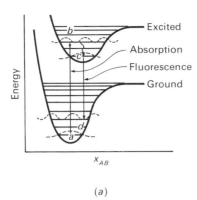

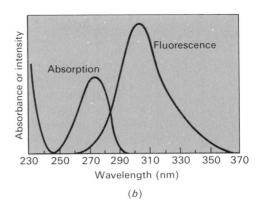

Figure 8.13 (*a*) Fluorescence arising from absorption by a bonding electron. The figure is similar to Figure 8.2, except that the forms of vibrational wave functions are shown (dashed lines). The height of the dashed line above its limiting values indicates the probability of different internuclear distances. (*b*) The excitation and emission spectra for the fluorescence of tyrosine.

energy curve for the excited state is such that the third vibrational level in this state will be favored.† Two things may now happen: (1) all of the excitation energy may be lost in a nonradiative manner, or (2) the molecule may lose some energy as heat and thus reach the lowest vibrational level of the excited state and then reradiate (line *cd*). If this happens, the emitted light will be of lower frequency (and thus longer wavelength) than the exciting light, and fluorescence will be observed.

It is commonly found that the wavelength of maximum fluorescence is relatively independent of the wavelength of the exciting light. This can be explained by the fact that the time required for the loss of excess vibrational energy in the excited state (about 10^{-12} sec) is very much less than the lifetime of the excited electronic state (about 10^{-9} sec). Thus excited molecules have plenty of time to reach the lowest vibrational level before they reemit, and the form of the fluorescence spectrum will be dictated by the relative probabilities of falling to the various vibrational levels in the *ground state*. Furthermore, transfer of electronic excitation within a given chromophore is a rapid process; the result is usually that excitation to any electronic level is followed by fluorescence from the *lowest* excited electronic state. The fluorescence spectrum thus serves, like the absorption spectrum, as a specific fingerprint for a compound. This, together with the fact that very small amounts of *emitted* radiation can be detected, makes fluorescence an exceedingly sensitive analytical tool. Figure 8.13(*b*) shows the typical fluorescence spectrum of tyrosine.

†For a fuller understanding of this point, some knowledge of the forms of the vibrational wave function is needed. See, for example, W. J. Moore, *Physical Chemistry*, 4th ed., Prentice-Hall, Inc., Englewood Cliffs, N.J., 1972.

In our discussion of fluorescence so far, we have assumed that the chromo-phore that emmitted the fluorescent radiation was the same group as the absorber. This need not be the case; under favorable circumstances excitation energy can be transferred from one chromophore to another. The requirements are (1) the possibility of dipole–dipole interaction between the chromophores and (2) an appreciable overlap of the fluorescence spectrum of the donor and the absorption spectrum of the acceptor. The fact that the transfer appears to operate by a dipole interaction mechanism explains the strong dependence of such transfer on the distance between the participating groups. The interaction is found to vary with the inverse sixth power of the separation r. The efficiency of transfer is given by the formula

$$\text{efficiency} = \frac{1}{1 + (r/R_0)^6} \tag{8.8}$$

Here R_0 is a distance parameter characteristic of the donor–acceptor pair and the medium between them. According to Equation (8.8), the efficiency increases from very small values to near unity if r becomes less than R_0. Since R_0 is ordinarily found to be of the order of 20 Å, the observation of such sensitized fluorescence thus serves as a useful yardstick for the distances between groups in macromolecules. The fact that the tyrosine and tryptophan groups in proteins usually satisfy both requirements for transfer (proximity and spectral overlap) explains an otherwise mystifying observation. Even if excitation is in the tyrosine band, the observed fluorescence is usually from tryptophan. This is because the excitation energy is transferred to tryptophan, which has the lowest-lying excited state (longest wavelength absorption) of the three. Thus energy transfer should be thought of as a common rather than as an unusual phenomenon.

Another mode of deexcitation, called *phosphorescence*, deserves mention. In phosphorescence, the molecule transfers to a long-lived excited state. Such states are *triplet* states, that is, electronic states in which the two electrons in an orbital have their spins in the same direction. Transitions between triplet states and the normally singlet (paired-electron) ground states are "forbidden" by quantum-mechanical selection rules. In practice this means that the rate of dropping back to the ground state is very low, thus accounting for the long life of the excited state. Of course, at room temperature radiationless transfer is much more probable than phosphorescent emission, so that phosphorescence is most commonly observed at very low temperatures.

Since singlet–triplet transitions are strongly forbidden as well, we must ask how the triplet state is reached in the first place. Two mechanisms are common: (1) the molecule may be raised to a singlet excited state in the ordinary fashion and then cross over to a triplet state of about the same energy, or (2) a chemical reaction may leave the products in such an electronic state.

All of the alternative modes of deexcitation mentioned above—radiationless transfer, excitation transfer, and crossing over to a triplet state—will play roles in determining the *quantum yield* of fluorescence. The quantum yield, Q, is defined as the ratio of photons emitted in fluorescence to photons absorbed. Obviously, it is a number that must vary between zero and 1. Changes in the quantum yield are often diagnostic of changes in the molecular environment of a chromophore, and can be utilized as a sensitive test for such variation. To take a simple example: the quantum yield of a dye will often increase when the dye molecule is taken up from solution and bound to a macromolecule; this provides a simple way to measure binding. Similarly, it is frequently found that the quantum yield of an aromatic amino acid changes when the protein of which it is a part is denatured.

8.6 Fluorescence Anisotropy

If the light used to excite fluorescence is plane polarized, absorption will be most probable for those molecules that happen to lie with their transition moments parallel to the plane of polarization; Equation (8.6) will apply. The state of polarization of the *emitted* light, however, will depend on a number of factors, including the orientation of the emitting dipole, and the rapidity of molecular rotation compared to the fluorescence lifetime. In most cases, the fluorescent light will be partially or wholly *depolarized*. We describe this in terms of a quantity called the *fluorescence anisotropy*

$$A = \frac{I_{\parallel} - I_{\perp}}{I_{\parallel} + 2I_{\perp}} \tag{8.9}$$

Here $I_{\parallel}$ and $I_{\perp}$ are the components of the emitted light intensity in planes parallel and perpendicular, respectively, to the polarization of the exciting light.† The principle of measurement is shown in Figure 8.14.

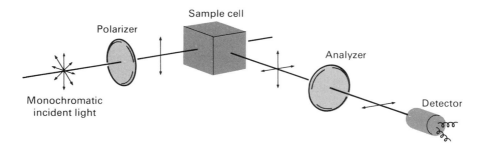

Figure 8.14 Measurement of fluorescence depolarization. The fluorescent light is detected at 90 deg through an analyzer. The analyzer can be set to measure either $I_{\parallel}$ or $I_{\perp}$.

†In the older literature, a quantity P, the *polarization*, was often used, with $P = (I_{\parallel} - I_{\perp})/(I_{\parallel} + I_{\perp})$. This is related to the anisotropy by $A = 2P/(3 - P)$.

Consider two limiting cases:

1. Molecular rotation is so slow that the chromophores cannot reorient significantly during the lifetime, τ_F, of the excited state. In this case, the anisotropy will have a maximum value, A_0. This will not usually approach the theoretical limit, $A_0 = 1$, since the directions of ground-state and excited-state transition moments will generally differ. Such depolarization in the absence of molecular motion is often referred to as *intrinsic* depolarization.

2. If the molecules can randomly reorient during the fluorescent lifetime, the emitted light will be wholly depolarized ($I_{\parallel} = I_{\perp}$) and $A = 0$.

The majority of actual cases will lie between these two extremes. Most biological macromolecules can undergo perceptible reorientation during a typical excited-state lifetime (10^{-9} to 10^{-8} sec). Furthermore, even if the molecule as a whole is sluggish, particular fluorescent groups (i.e., tryptophan side chains) are often sufficiently free to execute rapid rotation. Therefore, these motions will add to the intrinsic depolarization described above.

The most direct way to examine this phenomenon is to excite a fluorescent solution with a very short light pulse (ca. 10^{-9} sec) and then to follow the anisotropy over a period of time. Those molecules that emit immediately after the pulse will not have had time to rotate; they will show an anisotropy value close to A_0. On the other hand, molecules that emit later will have had time to undergo rotational motion; therefore, the anisotropy will decrease. Not surprisingly, the decay will be (at least for simple systems) exponential:

$$A(t) = A_0 e^{-t/\tau} \tag{8.10}$$

where τ is the rotational relaxation time. As one might expect, τ is related to the rotational diffusion coefficient (see Chapter 6):

$$\tau = \frac{1}{6\Theta} \tag{8.11}$$

Thus measurement of the time decay of fluorescence anisotropy provides a very convenient way to measure rotational motion of macromolecules. Two qualifications are required: First, Equations (8.10) and (8.11) are strictly correct only for approximately spherical particles. For asymmetric molecules the situation becomes more complex since more relaxation times are involved. Second, it must be remembered that local motions of the chromophore can contribute to the fluorescence decay. Unless the fluorescent group is firmly attached, the rotation of the molecule as a whole is not being monitored.

In an older, but still frequently utilized method, fluorescence depolarization is measured by determining the *steady-state anisotropy*. If a sample is continually illuminated with polarized light, an average value of the anisotropy will be observed. The value depends on the ratio between the fluorescent lifetime, τ_F, and the relaxation time τ.

$$\frac{1}{\bar{A}} = \frac{1}{A_0}\left(1 + \frac{\tau_F}{\tau}\right) \tag{8.12}$$

If we substitute into this equation the relationship between τ and Θ [Equation (8.11)] and the value of Θ for a sphere [Equation (6.27)], we obtain

$$\frac{1}{\bar{A}} = \frac{1}{A_0}\left(1 + \frac{\tau_F k}{v_m}\frac{T}{\eta}\right) \tag{8.13}$$

where v_m is the molecular volume ($4/3\pi R^3$) and k is the Boltzmann constant. According to Equation (8.13), varying the quantity T/η should allow determination of the molecular volume, provided that τ_F is known, since a graph of $1/\bar{A}$ versus T/η should have a slope $\tau_F k/v_m$.

This will be, of course, a hydrated molecular volume, and the equation as written is applicable only to spherical particles. To vary T/η, recourse is usually made to some such device as adding sucrose to the solutions and/or changing the temperature over a limited range in which the macromolecule may be presumed to be stable.

8.7 Magnetic Resonance Methods

The techniques of nuclear magnetic resonance (NMR) and electron spin resonance (ESR) have become of enormous importance in a wide variety of biochemical studies. Both of these methods are founded on the same principle: A particle such as an electron or an atomic nucleus may possess a property which can be thought of as corresponding to the classical-mechanical concept of *spin*. A spinning charge will have a magnetic dipole moment, **M**. In the presence of an external magnetic field such a dipole will tend to orient with the field, just as the magnetic dipole of a compass needle orients in the earth's magnetic field [see Figure 8.15(a)]. But there is a fundamental difference; the magnetic dipole of a quantum-mechanical particle cannot take just *any* orientation with respect to the applied field. Rather, the component of the moment in the field direction is restricted to certain quantized values, given by

$$M_z = g\frac{eh}{4\pi mc}m_s = g\beta m_s \tag{8.14}$$

Here e is the electronic charge, h is Planck's constant, m is the particle mass, c is the velocity of light, and g is an empirical constant. Like β ($= eh/4\pi mc$), g will differ for different particles. Values are given for some nuclei in Table 8.2. The quantity m_s is a quantum number. If the particle has a spin quantum number I, then m_s is allowed to take only the values $I, I-1, I-2, \ldots, -I$. For example, electrons and protons each have $I = \frac{1}{2}$; therefore, only values of $m_s = +\frac{1}{2}$ or $-\frac{1}{2}$ are allowed for such particles. The orientations of the magnetic dipole corresponding to these values are shown in Figure 8.15(b). The deuterium nucleus has $I = 1$, allowing three values of m_s: 1, 0, and -1. The consequence of these quantized orientations is quantization of the allowed energy levels for spinning particles in magnetic fields. The energy of a magnetic dipole of magni-

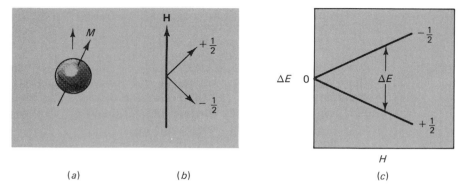

Figure 8.15 (*a*) A spinning charge generates a magnetic moment **M**. (*b*) The moment can take certain allowed orientations in a magnetic field, (*c*) The energy levels depend on the field strength.

TABLE 8.2   *Nuclei commonly used in biochemical NMR^a*

Isotope	Spin (I)	g	Abundance (%)	Relative Sensitivity	v at 10^4G^b (MHz)
^{1}H	$\frac{1}{2}$	5.585	99.98	(1.000)	42.57
^{2}H	1	0.857	0.02	0.0096	6.54
^{13}C	$\frac{1}{2}$	1.405	1.11	0.0159	10.70
^{19}F	$\frac{1}{2}$	5.257	100.00	0.834	40.05
^{31}P	$\frac{1}{2}$	2.263	100.00	0.0664	17.24

^{a}Adapted from D. Eisenberg and D. Crothers, *Physical Chemistry with Application to the Life Sciences*, Benjamin-Cummings Publishing Co., Menlo Park, Calif., 1979.
^{b}As predicted by Equation (8.20); 1 MHz = 10^6 sec^{-1}.

tude M is an external field of strength H is given by

$$E = -HM \cos \theta \qquad (8.15)$$

where θ is the angle between the field and dipole direction. Since $M \cos \theta$ is the component of the moment in the field direction,

$$E = -M_z H \qquad (8.16)$$

Thus, for the particle in Figure 8.15(*b*), the allowed energies are

$$E_{+1/2} = \frac{-g\beta H}{2} \qquad m_s = +\tfrac{1}{2} \qquad (8.17)$$

$$E_{-1/2} = \frac{g\beta H}{2} \qquad m_s = -\tfrac{1}{2} \qquad (8.18)$$

The energy difference between the two levels is therefore

$$\Delta E = g\beta H \tag{8.19}$$

Note that the energy-level difference is proportional to H, as is shown in Figure 8.15(c).

NMR and ESR spectrometry involve transitions between such energy levels. The frequency (v) for absorption must be given by $\Delta E = hv$, so

$$v = \frac{g\beta H}{h} \tag{8.20}$$

For fields of the kind that can be produced by large electromagnets ($H \simeq 10^4$ G), one finds that the frequencies v are in the megahertz† region, corresponding to the microwave region of the spectrum. In both NMR and ESR spectroscopy, a sample is placed in a strong magnetic field and excited with a microwave field. When the field frequency satisfies Equation (8.20), energy is absorbed. In classical-mechanical terms, the microwave field is in *resonance* with the oscillator. Note that there are two ways in which resonance can be achieved. A fixed external field, H, can be imposed, and the microwave frequency varied. Alternatively, a fixed frequency can be used, and the magnetic field H varied until resonance is obtained. With this general introduction, we now turn to specific features and applications of NMR and ESR.

NUCLEAR MAGNETIC RESONANCE

The use of NMR techniques is limited, of course, to nuclei that have nonzero spin. For biochemical applications, this means that most research utilizes the nuclei listed in Table 8.2. Although the naturally occurring isotope of nitrogen (^{14}N) has a nuclear spin of 1, the sensitivity is so very low and the absorption bands are so broad that little use has been made thereof. Some isotopes listed in Table 8.2 (such as ^{2}H and ^{13}C) are normally present in only very small amounts in biological samples, so substances are often enriched in these isotopes if their NMR is to be studied. Similarly, although fluorine is not a common element in biological materials, ^{19}F can sometimes be inserted into a compound to provide a "probe" at a specific site.

Equation (8.20) predicts that each isotope will yield a single resonance frequency (see Table 8.2). Were this rigorously true, NMR would have little attraction as a technique for biochemical study, for all of the protons in a molecule, for example, would show the same resonance line. But, in fact, they do not, for different protons are in different environments. They are covalently bonded to different atoms, and the various protons have a diversity of surroundings in a complex molecule. The applied field produces electron displacements in these different environments, which will generate magnetic fields, generally opposing the external field. Thus different nuclei in a molecule "see" slightly different

†One megahertz (MHz) = 10^6 cycles per second.

magnetic fields. These effects give rise to *chemical shifts* in NMR frequencies. The chemical shift, δ, is usually expressed in terms of the differences in applied field needed to produce resonance, relative to some reference material:

$$\delta = \frac{H_{\text{ref}} - H}{H_{\text{ref}}} \times 10^6 \qquad (8.21)$$

The factor 10^6 is included since chemical shifts are usually expressed in parts per million (ppm). A number of reference materials have been employed. For most proton NMR in aqueous solution, the methyl protons in 2,2-dimethyl-2-silapentane-5-sulfonate are now used; for nonaqueous solution the related tetramethyl silane is often employed. Values for chemical shifts of protons range from almost zero (as in CH_4) to about 16 ppm (in some enols). Thus differently located protons can often be identified by their chemical shifts, and their individual behavior studied.

An example is shown in Figure 8.16, which depicts the proton resonance spectrum of the base, purine, in aqueous solution. Assignment of the individual

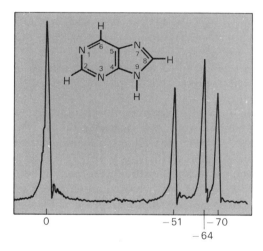

Figure 8.16 The nmr spectrum of purine in 0.8 *M* aqueous solution. The four lines from left to right are (1) CH_3Cl protons (an external reference), (2) H_8, (3) H_2, and (4) H_6. The N-9 proton is not observed due to exchange. After M. P. Schweitzer, S. I. Chan, G. K. Helmkamp, and P. O. P. Tso, *J. Am. Chem. Soc.*, **86**, 696 (1964). Copyright © 1964 by the American Chemical Society.

protons to particular bands was achieved by selectively deuterating different sites, and observing which lines were removed from the NMR spectrum. This is a common, and highly reliable method for assignment. Figure 8.17 exhibits a somewhat unusual feature of the proton NMR of purine in aqueous solution. It is found that purine exhibits an unusually large dependence of the δ values on concentration. This can be explained by the formation of stacked aggregates of such molecules in concentrated solutions. The magnetic fields generated by circulating electrons in the conjugated ring systems of adjacent molecules produce an additional chemical shift. This occurs in the upfield direction, and thus serves to decrease the normal downfield shift from the reference. In this

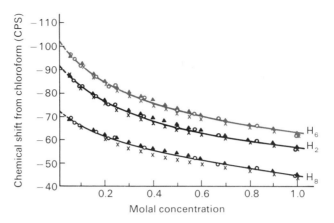

Figure 8.17 The concentration dependence of the chemical shifts observed in Figure 8.16. These results are consistent with a stacking of the purine molecules in aqueous solution: ○, experimental values; × and ▲, calculated values from two models of the stacking process. See S. I. Chan, M. P. Schweizer, P. O. P. Tso, and G. K. Helmkamp, *J. Am. Chem. Soc.*, **86**, 4182 (1964). Copyright © 1964 by the American Chemical Society.

case, then, NMR is able to provide information about both molecular structure and intermolecular interactions.

The application of proton NMR to the study of complex molecules such as proteins and nucleic acids has awaited the development of NMR spectrometers with magnets capable of producing very large fields. This is because such large fields spread out the field or frequency range corresponding to the range of chemical shifts usually encountered. For example, a "60-MHz" spectrometer will produce a field under which a 10-ppm range of δ corresponds to only 600 Hz, whereas a "360-MHz" instrument will spread the same lines over 3600 Hz. With such resolution, the proton resonances of individual groups in a protein can often be resolved.

NMR spectroscopy can also provide valuable information concerning kinetic processes. The subject is a complicated one, and cannot be treated in detail here, but a simple example can provide some intuitive understanding. Consider a nucleus which can exist in two different environments, because it is a part of a molecule undergoing an isomerization reaction ($A \rightleftharpoons B$). Now first let us suppose that the reaction is very slow. In this case, the NMR experiment will "see" two different kinds of molecules, and hence two resonance bands, with different chemical shifts, for the two states in which the nucleus may find itself [see Figure 8.18(*a*)]. On the other hand, if the interconversion reaction were speeded up to a very high rate, only the "average" environment for the nucleus would be observable, and the two bands would coalesce into a single band at the average chemical shift for the two states [Figure 8.18(*d*)]. For intermediate

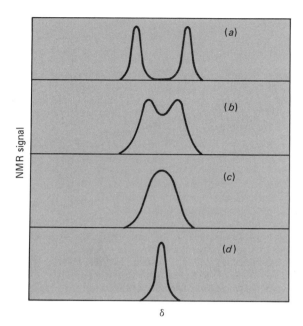

NMR signal

δ

Figure 8.18 The NMR signal for a nucleus that can exist in two interchangeable environments in a molecule. In example (*a*), the interconversion rate is slow, and the NMR line is split. In progressing from (*b*) ⟶ (*d*), more and more rapid inter-conversion is assured. For very rapid changes, a sharp line is obtained, (*d*), at a chemical shift determined by the fraction of time spent in each state.

states, patterns such as those shown in Figure 8.18(*b*) and (*c*) would be seen. With increasing reaction velocity, the two bands fuse into a single broad band, which then further narrows. It is clear from this example that NMR is sensitive to the kinetics of such processes. But more information is available. Just as a fast chemical process can "average" the environment, so can rapid molecular tumbling. In fact, it can be shown that magnetic resonance bands are sharper the more rapid the rotational motion of molecules. Large rigid molecules such as proteins or DNA turn slowly, and therefore exhibit broad NMR lines for their nuclei. If such molecules are denatured, so that they become random coil structures in which rapid motions of individual atoms become possible, the NMR bands corresponding to those nuclei become sharper. An example is shown in Figure 8.19, which depicts a part of the ^{13}C NMR spectrum of the synthetic polypeptide, poly-γ-benzyl-L-glutamate. Note that in the α-helical form the lines are broadened, and become even broader for higher-molecular-weight, more slowly tumbling helices. On the other hand, the resonances are sharp and independent of molecular weight in the random-coil forms, reflecting the freedom of motion in the backbone (α-carbon) and side-chain carbons (β, γ, benzyl) under these conditions.

Clearly, NMR possesses enormous advantages over all other methods for the study of molecular conformations in solution. The behavior of *individual* groups and atoms can be studied, even in a complicated macromolecule.

We have barely touched on the intricacies of NMR in this section. For

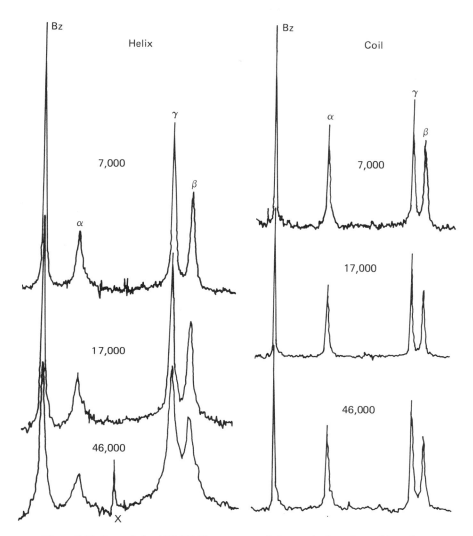

Figure 8.19 Part of the ^{13}C NMR spectrum of the synthetic polypeptide poly-γ-benzyl-L-glutamate. Spectra are shown for both the α-helical (*left*) and random coil (*right*) forms of this polymer. Resonances corresponding to the α, β, γ, and benzyl carbons are visible. Also, spectra for three different samples of molecular weights 7000, 17,000, and 46,000 are shown in each case. The line broadening in the helix spectra is due to the rigidity of structure and slowness of rotational motion of these rodlike molecules. After A. Allerhand and E. Oldfield, *Biochemistry*, **12**, 3428–3433 (1973). Reproduced with permission of authors and American Chemical Society.

more details, the reader is referred to the references listed at the end of the chapter.

ELECTRON SPIN RESONANCE

The basic principles of ESR spectroscopy are essentially the same as those of NMR. Equation (8.14) applies equally well to an unpaired electron:

$$M_z = g_e \frac{eh}{4\pi m_e c} m_s \tag{8.22}$$

where m_e is the electron mass. Since the electron spin quantum number m_s can take only the values $\pm\frac{1}{2}$, we have, by analogy to Equation (8.20),

$$\nu = \frac{g_e \beta_e H}{h} \tag{8.23}$$

for the electron spin resonance frequency. The constant β_e is called the *Bohr magneton*, and g_e has a value very close to 2.00. As in NMR, resonance will occur in the microwave region of the spectrum if magnetic fields of the order of 10^4 G are employed.

There are two principal situations in which unpaired electrons may be encountered in biochemical systems. First, a number of transition metal ions (Cu^{2+}, for example) have electrons with unpaired spins. Such ions occur in a number of metalloenzymes, and the study of such proteins has been greatly aided by ESR spectroscopy. Second, free radicals with unpaired electrons occur as intermediates in some reactions, or may be introduced via "spin labeling" reagents. For the latter purpose, reagents such as nitroxides are frequently employed.

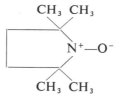

Just as with NMR, the most important applications of ESR spectroscopy arise from the ways in which the surroundings of an unpaired electron modify the spectrum. Interaction with the magnetic fields of neighboring nuclear spins will give rise to *hyperfine splitting* of the ESR bands. If a neighboring nucleus has a spin of $I = \frac{1}{2}$, its two allowed spin states can each interact with each of the two spin states of the electron. As a consequence, four states of different energy level will be produced. However, quantum-mechanical selection rules allow only two transitions instead of the four that might be expected. A neighboring nucleus such as ^{14}N, which has $I = 1$, allows for a more complex hyperfine splitting (see Figure 8.20). From such details of the ESR spectrum, information about the molecular environment of an unpaired electron can often be deduced. Furthermore, just as in NMR spectroscopy, the breadth of ESR bands reflects the free-

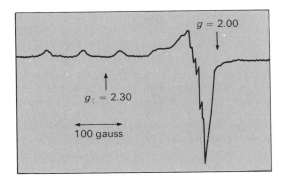

Figure 8.20 The ESR spectrum of a copper-conalbumin complex. As is usually the case, this has been recorded as a derivative spectrum. Two kinds of splittings can be seen. To the left (low field) are peaks due to interaction of the *d* electron with the spinning Cu nucleus. In the larger peak to the right are hyperfine splittings resulting from interactions with *N* ligands. After J. J. Windle, A. K. Wiersema, J. R. Clark, and R. E. Feeney, *Biochemistry*, **2**, 1341 (1963). Copyright © 1963 by the American Chemical Society.

dom of motion of the group containing the unpaired electron. This phenomenon has been used extensively in the study of membranes. Spin labels introduced into membranes show broad ESR signals at low temperatures, where the membrane is relatively rigid. As the temperature is raised and the membrane becomes fluid, the transition is accompanied by a marked narrowing of the ESR signal.

PROBLEMS

1. Cytosine has a molar extinction coefficient of 6×10^3 at 270 nm at pH 7. Calculate the absorbance and percent transmission of 1×10^{-4} and 1×10^{-3} *M* cytosine solutions in a 1-mm cell.

2. A protein with extinction coefficient $E^{1\%} = 16$ yields an absorbance of 0.73 when measured in a 0.5-cm cell. Calculate the weight concentration.

3. Consider two molecules A and B that have distinguishable but overlapping absorption spectra. Outline a mathematical method, utilizing absorption at two different wavelengths, for determining the amounts of A and B in a mixture. What will determine the choice of wavelengths?

4. The exciton splitting in polynucleotides appears to be of the order of 5 nm, with the absorption bands lying near 250 nm. Calculate, in kilocalories per mole, the corresponding interaction energies.

5. The following data describe the steady-state fluorescence polarization of a protein which has been labeled with a dye. The dye has a fluorescent lifetime of 7.0 nsec. Sucrose has been added to solutions at 20°C to increase η.

$T/\eta \times 10^{-4}$ (cgs)	$\bar{A}$
0.30	0.292
0.82	0.269
1.49	0.247
2.10	0.227
2.51	0.217
2.92 (H_2O)	0.206

(a)   Calculate A_0 and τ in H_2O at 20°C from these data.

(b)   Calculate the molecular volume. The protein has a polypeptide chain weight of 25,000, and $\bar{v} = 0.72$. What can you conclude about the protein under these conditions?

6.   It is found that there are two sites for attachment of fluorescent labels to the protein described in Problem 5. A pair is used for which R_0 is 23 Å. The energy transfer efficiency is found to be about 1.5%. Estimate the distance between the lables.

7.   Using the older quantity, $\bar{P} = (I_\parallel - I_\perp)/(I_\parallel + I_\perp)$, rewrite Equation (8.13) in terms of $\bar{P}$. This is the equation one will find in older texts and papers. Note that $\bar{A} = 2\bar{P}/(3 - \bar{P})$.

REFERENCES

General

Cantor, C. R. and P. R. Schimmel: *Biophysical Chemistry*, W. H. Freeman and Company, Publishers, San Francisco, 1980, Chaps. 7–9. A rigorous and quite detailed discussion of a number of kinds of spectroscopy.

Special Techniques

Chen, R. F. and H. Edelhoch (eds.): *Biochemical Fluorescence: Concepts*, Marcel Dekker, Inc., New York, Vol. 1, 1975; Vol. 2, 1976. A collection of chapters on special topics, some of very high quality.

Gadian, D. G.: *Nuclear Magnetic Resonance and Its Application to Living Systems*, Clarendon Press, Oxford, 1982. A small book that covers a remarkable range of topics.

Knowles, P. F., D. Marsh, and H. W. E. Rattle: *Magnetic Resonance of Biomolecules*, John Wiley & Sons, Inc., New York, 1976. Covers both NMR and ESR fundamentals and applications to biochemistry. Excellent.

Pesce, A. J., C. J. Rosen, and T. L. Pasby: *Fluorescence Spectroscopy*, Marcel Dekker, Inc., New York, 1971. A good general review, but somewhat dated.

Susi, H.: "Infrared Spectra of Biological Macromolecules and Related Systems," in *Structure and Stability of Biological Macromolecules* (S. N. Timasheff and G. Fasman eds.), Marcel Dekker, Inc., New York, 1969, pp. 576–663.

Chapter Nine

SCATTERING

Absorption is not the only manner in which molecules can interact with radiation. Suppose that we illuminate a solution with light of a wavelength that is far from any absorption band. Molecules are polarizable; their distribution of electronic charge can be shifted by an electric field. The classic description pictures radiation as a sinusoidally varying electromagnetic field. This should produce a sinusoidal oscillation of the electrons within a molecule. *Such oscillating charges will cause each molecule to act as a minute antenna, dispersing some of the energy in directions other than the direction of the incident radiation.* This is the basis for all scattering phenomena.

9.1 Light Scattering: Fundamental Concepts

Since Chapters 9 through 11 will be concerned primarily with the various consequences of scattering, we should look at the process in detail. Consider the simplest situation, the scattering from a single molecule. We shall make the light monochromatic (by filters, for example, or use a laser) and polarized, although we shall remove the latter restriction later. The light is directed along the x axis in Figure 9.1 and polarized with electric vector in the z direction. The molecule is at $x = 0$, $y = 0$, and $z = 0$. We assume λ, the wavelength, to be so long that we can put the molecule *at* the origin, without having to worry about its size.

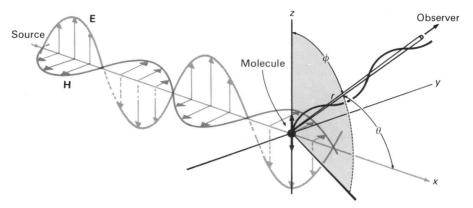

Figure 9.1 Scattering of radiation by a particle. Only the electric vector of the radiation scattered in the particular direction given by θ and ϕ is shown. Radiation scattered in this direction is polarized in the plane defined by the z axis and r.

Later, we shall consider what happens if there are many molecules in the scattering sample, or if the molecule is very large, or if the wavelength of the radiation is very small.

Since absorption phenomena are not to be considered, a classical rather than a quantum-mechanical treatment will be appropriate. The incident radiation can be described in terms of the electric field

$$\mathbf{E} = \mathbf{E}_0 \cos 2\pi v \left(t - \frac{x}{c} \right) \tag{9.1}$$

and the perpendicular magnetic field

$$\mathbf{H} = \mathbf{H}_0 \cos 2\pi v \left(t - \frac{x}{c} \right) \tag{9.2}$$

In these equations, v is frequency, t is time, and c is the velocity of light. For the present discussion, only the electric field is of importance.† The electric field at the molecule ($x = 0$) varies with time as

$$\mathbf{E} = \mathbf{E}_0 \cos 2\pi v t \tag{9.3}$$

Since all molecules are polarizable, this field will produce an oscillating dipole moment

$$\boldsymbol{\mu} = \alpha \mathbf{E} = \alpha \mathbf{E}_0 \cos 2\pi v t \tag{9.4}$$

where α is the molecular polarizability. If the molecule is isotropic, the dipole moment will be in the direction of the electric vector, that is, along the z axis. This oscillating moment will act as a source of radiation; we may define the radiated field in terms of the coordinates shown in Figure 9.1. The amplitude

---

†The magnetic field and its interactions with electrons will be important in the discussion of optical rotatory phenomena (see Chapter 10).

of the electric field produced by an oscillating dipole, at a distance r from the dipole and at an angle ϕ with respect to the direction of polarization (the z axis), is given by electromagnetic theory[†]:

$$E_r = \left[\frac{\alpha E_0 4\pi^2 \sin \phi}{r\lambda^2} \right] \cos 2\pi v \left(t - \frac{r}{c} \right) \qquad (9.5)$$

The term in brackets in Equation (9.5) is the part of interest to us. It represents the amplitude of the scattered wave. The intensity of the radiation (the energy flow per square centimeter) depends on the square of the amplitude. We wish to compare the intensity i of the scattered radiation to the intensity I_0 of the incident radiation. The latter is proportional to the square of *its* amplitude, E_0:

$$\frac{i}{I_0} = \frac{(\alpha E_0 4\pi^2 \sin \phi / r\lambda^2)^2}{E_0^2} = \frac{16\pi^4 \alpha^2 \sin^2 \phi}{r^2 \lambda^4} \qquad (9.6)$$

This equation tells a great deal about the scattered light. Its intensity falls off with r^{-2}, as radiation from a point source must. The intensity of scattering increases rapidly with decreasing wavelength.[‡] The intensity depends on the angle ϕ; there is no radiation along the direction in which the dipole vibrates. A graph, in polar coordinates, of the radiation intensity looks like the doughnut-like surface in Figure 9.2(*a*).

Normally, unpolarized radiation is used in light-scattering experiments. Since this may be regarded as a superposition of many independent waves, polarized in random directions in the *yz* plane, we may imagine that the resulting scattering surface would correspond to the addition of surfaces such as that shown in Figure 9.2(*a*), rotated at random with respect to one another. The resulting surface is seen in Figure 9.2(*b*); it looks somewhat like a dumbbell, with the narrowest part in the *yz* plane. The equation for scattering of unpolarized radiation can be easily obtained[§]; it differs from Equation (9.6) only in that $(1 + \cos^2 \theta)/2$ is substituted for $\sin^2 \phi$:

$$\frac{i}{I_0} = \frac{8\pi^4 \alpha^2}{r^2 \lambda^4} (1 + \cos^2 \theta) \qquad (9.7)$$

Here θ is the angle between the incident beam and the direction of observation. The distribution of scattering intensity in the *xy* plane is shown in Figure 9.2(*c*); evidently, the scattering is symmetrical in forward and backward directions.

[†]The electromagnetic field surrounding a moving accelerating charge [the displaced electron(s)] will consist of three parts: the static field, depending only on the magnitude of the charge; the induction field, depending on its velocity; and the radiation field, depending on the acceleration of the charge. Only the last concerns us here, for the first two fall off rapidly with distance from the charge [see Feynman et al. (1963)].

[‡]The strong dependence of scattering on λ accounts for the blue of the sky, since we observe the sunlight scattered by the air and its contaminants, and the blue light is scattered more than red.

[§]See Tanford (1961).

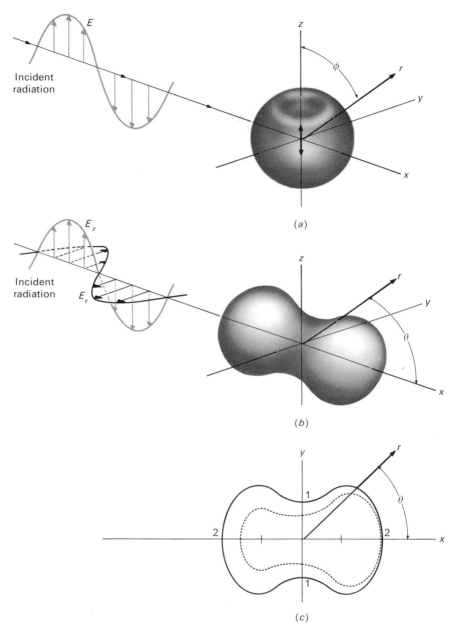

Figure 9.2 (*a*) Distribution of intensity of scattering of incident light polarized in the *z* direction. In this figure, and in (*b*), the distance from the origin to the surface along a direction *r* shows the intensity in that direction. (*b*) Distribution of scattering intensity for unpolarized incident light. (*c*) A section in the *xy* plane of the surface of (*b*). The solid line shows the intensity of scattering as a function of θ for Rayleigh scattering (small particle). The dotted line is for a larger particle.

Equation (9.7) provides a completely adequate description of the scattering of light by a single small, isotropic particle. However, as it stands, Equation (9.7) is not of much use to the biochemist, who wishes to study solutions and to use the measurement of scattering to find out something about the solute molecules. We must therefore consider the situations that arise when the scattering is by a collection of particles.

9.2 Scattering from a Number of Particles

To see the kind of thing that happens when scattering by more than one particle is occurring, let us consider the simplest case: two particles. Imagine, as in Figure 9.3, two identical scattering centers, fixed in space and illuminated by a common source. The passing electromagnetic waves stimulate each center to

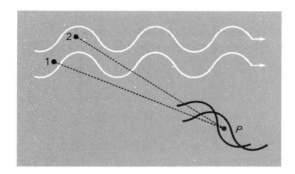

Figure 9.3 Scattering from a pair of atoms or molecules. The point of observation, *P*, is assumed to be far from the scatterers.

scatter. We observe the scattering at some distant point *P*. What is the amplitude and intensity of the scattered radiation at *P*? In general, there will be a phase difference in the scattered waves received at *P* from 1 and 2. There will be two reasons for this: First, the centers 1 and 2 lie, at any instant, at different positions in the incident field—they will scatter out of phase. A further phase difference can result from the different distances of *P* from points 1 and 2. To keep the expressions as simple as possible, we can represent the electric fields seen at *P* by the expression

$$A \cos (2\pi v t + \phi_1) \qquad \text{from center 1}$$

and

$$A \cos (2\pi v t + \phi_2) \qquad \text{from center 2}$$

(9.8)

Here *A* is the amplitude of the wave scattered by each particle, and ϕ_1 and ϕ_2 are the phases of the scattered waves, relative to some arbitrary reference. The total electric field seen at *P* will then be

$$E = A \cos (2\pi v t + \phi_1) + A \cos (2\pi v t + \phi_2) \qquad (9.9)$$

We may rewrite this in a more useful form by using the well-known identity:

$\cos x + \cos y = 2 \cos \frac{1}{2}(x - y) \cos \frac{1}{2}(x + y)$:

$$E = \underbrace{2A \cos \tfrac{1}{2}(\phi_1 - \phi_2)}_{\text{amplitude}} \cos \left(2\pi vt + \frac{\phi_1 + \phi_2}{2}\right) \qquad (9.10)$$

This is an expression for the electromagnetic wave produced at P by the super-position of the two scattered waves. Its amplitude depends on the phase differ-ence $\phi_1 - \phi_2$. If the scattered waves are 180° out of phase, the amplitude is zero; the waves destructively interfere. If the phase difference is zero, or some multiple of 360°, the waves reinforce. We can calculate the intensity of the combined wave by taking the square of the amplitude:

$$i_{1,2} = 4A^2 \cos^2 \left(\frac{\Delta\phi}{2}\right) \qquad (9.11)$$

or, using another standard identity,

$$i_{1,2} = 2A^2(1 + \cos \Delta\phi)$$
$$= 2i(1 + \cos \Delta\phi) \qquad (9.12)$$

where i is the intensity scattered by one particle. Equation (9.12) states that the intensity we shall see from a pair of scatterers depends on the direction from which we observe them, ranging† from 0 to $4i$.

The situation then is as follows: For given locations of two scatterers, the intensity of scattering will be greater or less than the sum of the individual intensities, depending on the direction of observation. Now let us suppose that the two scatterers are free to move at random with respect to one another. In this case, all values of $\Delta\phi$ will be equally probable over a period of time. Then the time-averaged cosine of $\Delta\phi$ will be zero, and $i_{1,2} = 2i$. This may easily be generalized to a result for n scatterers; if they are in random motion with respect to one another, the total intensity of scattering will be the sum of the individual scattering intensities. This result obviously applies to the scattering by an ideal gas. However, let us turn to the behavior of condensed phases.

Consider, as in Figure 9.4, the scattering from a perfect crystal. It is easy to show, from the arguments above, that the scattering of long-wavelength radia-tion (long, as compared to the lattice spacing) will be zero, except in the forward direction. Consider any one scatterer i, observed from a given point, P. If the crystal is large enough and the lattice spacing small enough, there will always be another scattering center at some point j, which is so located that the phases of the scattered waves differ by some odd multiple of 180°. In other words, the crystal can be considered as made up of scattering elements that cancel one another. The only exception to this rule will be for scattering in the direction taken by the incident beam. Here the scattered waves will always be in phase, and reinforcement must occur. For a pure liquid, the situation will be similar.

†The intensity averaged over all directions is, of course, $2i$, corresponding to the fact that the energy lost to the incident beam is twice that for a single scatterer.

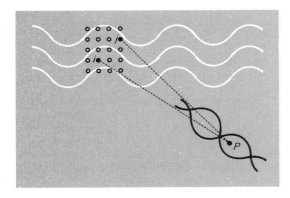

Figure 9.4 Scattering from a perfect crystal. If the crystal is large enough and the interparticle spacing small enough, there will be, for every scatterer, *i*, another *j* that is so placed that at *P* their scattered waves exactly cancel.

We can consider dividing the liquid into cells much smaller than λ (Figure 9.5). Each of these cells may be considered a scattering center. If they were all of equal density, and thus of equal scattering power, the argument for a crystal would apply: no scattering except in the direction of the beam. But in a real liquid, *fluctuations* in density occur from instant to instant; thus the densities in a pair of cells that are so located as to scatter out of phase will not in general be quite equal. Hence their scattering will not quite cancel. Thus, because of local fluctuations in density, a pure liquid will scatter somewhat more than a perfect crystal, though not nearly so much as the sum of the individual molecular scattering powers. Of course, once again the scattering in the forward direction will be undiminished, since phase differences are zero.

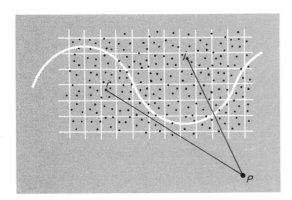

Figure 9.5 Scattering from a liquid. The fluid has been divided into cells small compared to λ. Each can contain a number of molecules. Each cell is considered to be a scattering center. Suppose *i* and *j* are so placed that their scattering at point *P* is out of phase (as in Figure 9.4). If *i* and *j* contained equal numbers of molecules (had equal scattering power), their scattering would cancel at *P*. But fluctuations in density prevent complete cancellation from occurring. Thus some light is scattered.

9.3 Rayleigh Scattering from Solutions of Macromolecules

Dissolved macromolecules scatter light quite strongly, and this phenomenon can be used to measure their molecular weights, and in some cases their dimensions as well. In this section we shall derive equations for the scattering from a solution of macromolecules whose dimensions are small compared to the wavelength of the light. This is referred to as *Rayleigh scattering*. We shall be interested in the excess scattering produced by the solute molecules; it will be assumed that solvent scattering has been subtracted out.

It must be emphasized that the excess scattering observed from a solution arises from the fact that the molecules are nonuniformly distributed in the solution. Different small elements of volume will contain, at any instant, different numbers of solute molecules in them. Thus the scattering from such elements does not entirely cancel. Furthermore, there will be continual fluctuations in the numbers of molecules in each volume element. There are thus two kinds of scattering experiments. In the *dynamic* scattering experiment, as described in Chapter 4, attention is paid to the kinetics of the fluctuations in the scattered light intensity. But these fluctuation ripples will be superimposed on a background of scattering which arises from the fact that at any instant the concentration in one volume element is unlikely to be the same as in another. It is this steady-state scattering that we are concerned with here.

This average scattering can be calculated in the following way. We consider 1 cm^3 of the solution to be divided into N volume elements, each of volume v. Since we are interested in the excess scattering of solution over pure solvent, we can write Equation (9.7) for the instantaneous scattering from a single volume element as

$$\frac{i_v}{I_0} = \frac{8\pi^4 \alpha_v^2 (1 + \cos^2 \theta)}{r^2 \lambda^4} \tag{9.13}$$

where α_v is the *excess* polarizability of one volume element. It is the polarizability of that volume minus the polarizability of an equal volume of pure solvent. Since α_v fluctuates with fluctuating macromolecular concentration, we should write

$$\alpha_v^2 = (\overline{\alpha_v} + \delta\alpha_v)^2 = (\overline{\alpha_v})^2 + 2\overline{\alpha_v}\,\delta\alpha_v + (\delta\alpha_v)^2 \tag{9.14}$$

where $\overline{\alpha_v}$ is the average value of α_v, and $\delta\alpha_v$ is the fluctuating part. If we now wish to calculate the average contribution of this volume element to the total scattering, we must take the average value of α_v^2:

$$\overline{\alpha_v^2} = (\overline{\alpha_v})^2 + 2\overline{\alpha_v}\,\overline{\delta\alpha_v} + \overline{\delta\alpha_v^2} \tag{9.15}$$

But the first term will be the same for *all* volume elements; this term will contribute nothing to the total scattering, just as the volume elements in a perfect crystal contribute nothing. The second term also vanishes, for positive and negative values of $\delta\alpha_v$ will be equally probable ($\overline{\delta\alpha_v} = 0$). Thus we are left with only the third term, and Equation (9.13) reduces to

$$\frac{\overline{i_v}}{I_0} = \frac{8\pi^4 \overline{(\delta\alpha_v)^2}(1 + \cos^2 \theta)}{r^2 \lambda^4} \tag{9.16}$$

At this point, it is useful to relate the fluctuations in polarizability to the fluctuations in concentration (C) in the volume element. The polarizability, for radiation of optical frequencies, is related to the refractive index, n. For the excess polarizability, we have

$$n^2 - n_0^2 = 4\pi N \alpha_v \tag{9.17}$$

where n_0 is the refractive index of the solvent. There are N volume elements, each of excess polarizability α_v, in each cubic centimeter of solution. Since

$$\delta \alpha_v = \frac{d\alpha_v}{dC} \delta C \tag{9.18}$$

$$\overline{(\delta \alpha_v)^2} = \left(\frac{d\alpha_v}{dC}\right)^2 \overline{(\delta C)^2} \tag{9.19}$$

Differentiating Equation (9.17) with respect to C, we have

$$2n \frac{dn}{dC} = 4\pi N \frac{d\alpha_v}{dC} = \frac{4\pi}{v} \frac{d\alpha_v}{dC} \tag{9.20}$$

or

$$\frac{d\alpha_v}{dC} \simeq \frac{v n_0}{2\pi} \frac{dn}{dC} \tag{9.21}$$

where we have made use of the identity $N = 1/v$, and approximated n by n_0. So Equation (9.16) becomes

$$\frac{\bar{i}_v}{I_0} = \frac{2\pi^2 v^2 n_0^2 (dn/dC)^2 \overline{(\delta C)^2}(1 + \cos^2 \theta)}{r^2 \lambda^4} \tag{9.22}$$

Now we must evaluate the mean-square concentration fluctuation, $\overline{(\delta C)^2}$. We are dealing with small fluctuations about an equilibrium concentration, C. We can express the corresponding fluctuation in free energy in terms of a Taylor's series:

$$\delta G = \left(\frac{\partial G}{\partial C}\right)_{T,P} \delta C + \frac{1}{2}\left(\frac{\partial^2 G}{\partial C^2}\right)_{T,P} (\delta C)^2 + \cdots \tag{9.23}$$

Since G is a minimum at the equilibrium concentration, $(\partial G/\partial C)_{T,P} = 0$. Therefore,

$$\delta G \simeq \frac{1}{2}\frac{\partial^2 G}{\partial C^2}(\delta C)^2 \tag{9.24}$$

The *average* value of $(\delta C)^2$, which is what we seek, can be obtained from a Boltzmann distribution, with δG as the energy:

$$\overline{(\delta C)^2} = \frac{\int_0^\infty (\delta C)^2 e^{-\delta G/kT}\, d(\delta G)}{\int_0^\infty e^{-\delta G/kT}\, d(\delta G)} \tag{9.25}$$

Inserting Equation (9.24), the integrals may be easily evaluated to yield

$$\overline{(\delta C)^2} = \frac{kT}{(\partial^2 G/\partial C^2)_{T,P}} \tag{9.26}$$

Now, all that remains is to evaluate $(\partial^2 G/\partial C^2)_{T,P}$. If there are, on the average,

n_2 moles of solute of molecular weight M in the volume element v, we may write

$$\left(\frac{\partial G}{\partial C}\right)_{T,P} = \left(\frac{\partial G}{\partial n_2}\right)_{T,P}\frac{\partial n_2}{\partial C} = \mu_2\frac{v}{M} \tag{9.27}$$

since $n_2 = vC/M$. The quantity μ_2 is, of course, the chemical potential. Differentiating once again, we have

$$\left(\frac{\partial^2 G}{\partial C^2}\right)_{T,P} = \frac{v}{M}\left(\frac{\partial \mu_2}{\partial C}\right)_{T,P} \tag{9.28}$$

Since

$$\mu_2 = \mu_2^0 + RT\ln C + RT\ln y \tag{9.29}$$

we have

$$\left(\frac{\partial^2 G}{\partial C^2}\right)_{T,P} = \frac{v}{M}\frac{RT}{C}\left(1 + C\frac{\partial \ln y}{\partial C}\right) \tag{9.30}$$

If we now insert Equation (9.30) into (9.26), we find that

$$\overline{(\delta C^2)} = \frac{MC}{\mathfrak{N}v[1 + C(\partial \ln y/\partial C)]} \tag{9.31}$$

where $\mathfrak{N}$ is Avogadro's number, which equals R/k. We can now obtain the expression for the scattering per unit volume of solution by inserting Equation (9.31) into (9.22) and multiplying by $N = 1/v$, the number of elements in unit volume:

$$\frac{i}{I_0} = \frac{2\pi^2 n_0^2(dn/dC)^2(1 + \cos^2\theta)MC}{\mathfrak{N}r^2\lambda^4[1 + C(\partial \ln y/\partial C)]} \tag{9.32}$$

This equation can be somewhat simplified by grouping a number of experimental parameters together into a quantity called the *Rayleigh ratio*:

$$R_\theta = \frac{i}{I_0}\frac{r^2}{1 + \cos^2\theta} \tag{9.33}$$

whereas a number of the constants are usually combined as

$$K = \frac{2\pi^2 n_0^2(dn/dC)^2}{\mathfrak{N}\lambda^4} \tag{9.34}$$

With these definitions, Equation (9.32) may be abbreviated as

$$\frac{KC}{R_\theta} = \frac{1}{M}\left(1 + C\frac{\partial \ln y}{\partial C}\right) \tag{9.35}$$

Note that this thermodynamic derivation allows us to include the effects of solution nonideality. In terms of a virial expansion of $\ln y$ (Chapter 2), we have

$$\frac{KC}{R_\theta} = \frac{1}{M} + 2BC + \cdots \tag{9.36}$$

For an ideal solution, of course, Equations (9.35) and (9.36) both reduce to

$$\frac{KC}{R_\theta} = \frac{1}{M} \tag{9.37}$$

These equations show that measurement of Rayleigh scattering can be used to determine the absolute molecular weights of macromolecular solutes. Note that there are similarities between Equation (9.36) and the equations for the osmotic pressure of a nonideal solution. In both cases, an extrapolation of the experimentally measured quantities to zero concentration is needed to obtain the true molecular weight.

If the solute is heterogeneous, the light-scattering method must give an average molecular weight. In fact, it is the *weight average* that is obtained. This follows from the fact that at low concentration the total R_θ must be the sum of the contributions from the various solute components, $R_{\theta i}$:

$$R_\theta = \sum_i R_{\theta i} = \sum_i KC_i M_i = K \frac{\sum_i C_i M_i}{\sum_i C_i} C = K\bar{M}_w C \qquad (9.38)$$

where we have assumed K (and therefore dn/dC) to be the same for all components and have made use of the identity $\sum_i C_i = C$, where C is the total weight concentration.

Equation (9.36) indicates that to measure the molecular weight we must determine R_θ at a number of concentrations and extrapolate KC/R_θ to $C = 0$. The measurement of R_θ is usually carried out in a *photometer* similar to that diagrammed in Figure 9.6. The intensity of light scattered at a given angle (90°, for example) is compared with the intensity of the incident light. A complication arises from the fact that the solution must be scrupulously clean; small amounts of dust will contribute as extremely large molecules to the average in Equation (9.38), leading to serious errors. Therefore, solutions must be carefully filtered or centrifuged before use.

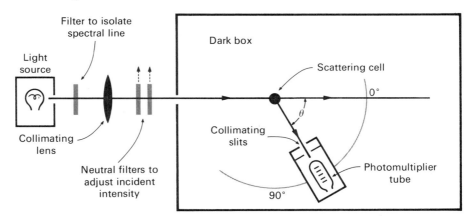

Figure 9.6 A schematic sketch of a light-scattering photometer. The photomultiplier is mounted on a turntable so that it can be rotated to observe at any angle θ. The incident beam intensity (reduced by neutral filters) is compared to the intensity scattered at various angles. In modern instruments a laser is often used as a light source to provide a monochromatic, well-collimated beam.

In addition to R_θ values, calculation of the molecular weight requires a knowledge of dn/dC. Since the difference in refractive index between a dilute solution and pure solvent is very small, a *differential* refractometer is commonly used.

9.4 Scattering by Larger Particles

So far, we have assumed that the greatest dimension of the scattering particle is small compared to the wavelength of the light. If this is not true, the light-scattering experiment becomes more interesting and informative. Examine Figure 9.7. The incident light is shown inducing dipole oscillation at a pair of points that are an appreciable fraction of a wavelength apart; the dipoles within a given molecule are oscillating out of phase. Since the scattering centers within a given large molecule are (more or less) fixed with respect to one another, we can no longer consider them to be independent scatterers. We must therefore, take into account interference between these scattering centers.

The calculation of this effect for a macromolecule must include the interference between light scattered from all pairs of scattering points within the molecule. Since we also have to consider all orientations of a molecule that is assumed to be in rapid Brownian movement the problem is complicated. The result, which is of general validity and will be encountered in a number of situations, may be stated as follows: We call $P(\theta)$ the ratio of the scattered intensity at an angle θ to that which would be observed if the particle had the same molecular weight (same number of scattering dipoles) but infinitesimal dimensions compared to λ. Then $P(\theta)$ for any aggregate of scatterers turns out to be†

$$P(\theta) = \frac{1}{N^2} \sum_{i=1}^{N} \sum_{j=1}^{N} \frac{\sin hR_{ij}}{hR_{ij}} \tag{9.39}$$

where N is the number of scattering centers within the molecules (we could take them to be the atoms), R_{ij} is the distance between any pair of centers i and j, and h is given by

$$h = \frac{4\pi}{\lambda} \sin \frac{\theta}{2} \tag{9.40}$$

where λ is the wavelength of light in the medium (that is, $\lambda = \lambda_0/n$, where n is the refractive index and λ_0 the wavelength in vacuum). Equation (9.39) is a double sum that extends over all pairs of scattering centers. To show the behavior of $P(\theta)$ in limiting cases, let us represent $(\sin x)/x$ by a Taylor's series. The familiar expansion for $\sin x$ at small x, when divided through by x, gives

†The derivation of this equation is beyond the scope of this book [see Tanford (1961)]. It might be worthwhile to do Problem 4 at this point to gain some idea of how Equation (9.39) is derived.

$$\frac{\sin hR_{ij}}{hR_{ij}} \simeq 1 - \frac{(hR_{ij})^2}{6} + \frac{(hR_{ij})^4}{120} + \cdots \tag{9.41}$$

This immediately yields several results. As either $h \rightarrow 0$ or $R_{ij} \rightarrow 0$, we have

$$P(\theta) \longrightarrow \frac{1}{N^2} \sum_{i=1}^{N} \sum_{j=1}^{N} (1) = \frac{1}{N^2} N^2 = 1 \tag{9.42}$$

Thus at very low angles, at very long wavelengths, or for very small particles, Rayleigh scattering is approached. The situation at $\theta = 0$, in fact, can be inferred from Figure 9.7; as usual, there is no phase difference in light scattered directly in the forward direction.

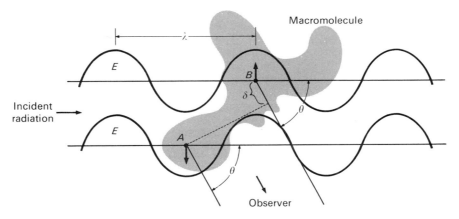

Figure 9.7 Scattering from a macromolecule that is large compared to λ. Two points from which scattering occurs are shown at A and B. The phase of the radiation (and hence of the induced dipoles) is clearly different. Also, the two points are at different distances from the observer.

At angles greater than zero, $P(\theta)$ will always be less than unity, which means that the scattering intensity at these angles will always be less for an extended particle than for a compact particle of the same weight. A typical graph of the angular dependence of light scattering for a large particle is shown as the dotted line in Figure 9.2(c). We have multiplied the $(1 + \cos^2 \theta)/2$ factor by $P(\theta)$.

It may appear that this effect of particle size is merely an exasperating complication, but, in fact, it allows us to determine particle dimensions from light scattering. Suppose that the particles are in such a size range that we need consider only the first two terms in Equation (9.41). Then

$$P(\theta) = \frac{1}{N^2} \sum_{i=1}^{N} \sum_{j=1}^{N} (1) - \frac{h^2}{6N^2} \sum_{i=1}^{N} \sum_{j=1}^{N} R_{ij}^2 \tag{9.43}$$

The first term in Equation (9.43) is unity, and the sum in the second has already been encountered in another form in Chapter 2. The *radius of gyration* of a particle, R_G, can be expressed in terms of the distances between pairs of mass units in the molecule:

$$R_G^2 = \frac{1}{2N^2} \sum_{i=1}^{N} \sum_{j=1}^{N} R_{ij}^2$$

Therefore,

$$P(\theta) = 1 - \frac{h^2 R_G^2}{3} + \cdots = 1 - \frac{16\pi^2 R_G^2}{3\lambda^2} \sin^2 \frac{\theta}{2} + \cdots \qquad (9.44)$$

There has been no assumption made about the shape of the particle in this derivation. Thus the angular dependence of light scattering can give us un-ambiguously the quantity R_G. This will only be useful if the particle is fairly large, as can be seen readily from Equation (9.44). If the radius of gyration is less than one fiftieth of the wavelength (that is, less than about 80 Å for visible light in water), the deviation of $P(\theta)$ from unity will not amount to more than a small percentage at any angle. Therefore, measurement of the angular dependence of the scattering of visible light will be useful only for the larger macromolecules.

Since the scattering for a real solution of large macromolecules depends on both angle and solute concentration, it is evident that R_θ must be measured as a function of both variables. If we recall that the definition of $P(\theta)$ says

R_θ(real particle) $= P(\theta) \times R_\theta$(same particle, if it showed Rayleigh scattering)

then Equation (9.36) must be rewritten as

$$\frac{KC}{R_\theta} = \frac{1}{P(\theta)}\left(\frac{1}{M} + 2BC\right) \qquad (9.45)$$

or, using the approximate form given in Equation (9.44),

$$\frac{KC}{R_\theta} = \frac{1}{1 - (16\pi^2 R_G^2/3\lambda^2)\sin^2(\theta/2)}\left(\frac{1}{M} + 2BC\right)$$

$$\simeq \left(1 + \frac{16\pi^2 R_G^2}{3\lambda^2}\sin^2\frac{\theta}{2}\right)\left(\frac{1}{M} + 2BC\right) \qquad (9.46)$$

In the case of heterogeneous samples, M is to be replaced by $\bar{M}_w$. Evidently, to obtain M it is necessary to extrapolate KC/R_θ to zero angle *and* zero concentration. In the method devised by Zimm (1948), both extrapolations are made on the same graph. Figure 9.8 shows a Zimm plot for data on the light scattering of a macromolecule. The quantity KC/R_θ has been graphed versus $\sin^2(\theta/2) + kC$, where k is an arbitrary constant, here taken to be 2000 to give a convenient scale. For each measurement of R_θ (at a particular value of θ and C), we measure a distance $\sin^2(\theta/2) + kC$ on the abscissa and plot the corresponding value of R_θ as the ordinate. Now consider a series of points at a given concentration but different angles. Such a series will lie along a line such as A. The value of KC/R_θ that will be approached at *zero angle* and *this concentration* is given by the point at which the line A has the abscissa value kC—the point x_A in this case. A series of such points have been obtained, one at each concentration. Together, they lie on a line (denoted by 0°) that represents the

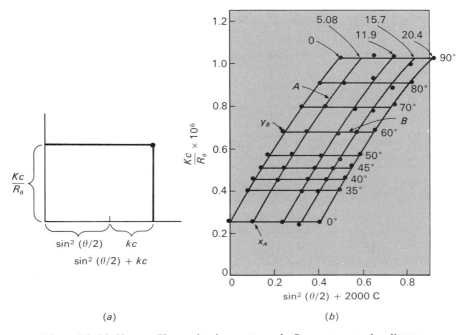

(a) (b)

Figure 9.8 (a) How a Zimm plot is constructed. One measures the distance $\sin^2(\theta/2) + kc$ on the abscissa, KC/R_θ on the ordinate. (b) Light scattering of DNA. The numbers on top refer to concentrations; those on the right to angles. Data from P. Doty and B. Bunce, *J. Am. Chem. Soc.*, **74**, 5029, (1952). Copyright © 1952 by American Chemical Society.

scattering of this sample at various concentrations and zero angle. The equation of this line is, from Equation (9.4b),

$$\left.\frac{KC}{R_\theta}\right|_{\theta=0} = \frac{1}{M} + 2BC \tag{9.47}$$

and thus one finds M and B from the intercept and slope.

Now suppose that instead of proceeding as above, we choose a series of points at a given scattering angle but different concentrations. If we extrapolated such a series such as the line B to the point where $C = 0$ [where $\sin^2(\theta/2) + kC = \sin^2(\theta/2)$], the point y_B would be obtained. This represents the value of KC/R_θ approached at a given θ as $C \to 0$. It is easy to see that the series of zero-concentration points (denoted by 0) is described by the equation

$$\left.\frac{KC}{R_\theta}\right|_{C=0} = \frac{1}{P(\theta)}\frac{1}{M} = \frac{1}{M}\left(1 + \frac{16\pi^2}{3}\frac{R_G^2}{\lambda^2}\sin^2\frac{\theta}{2}\right) \tag{9.48}$$

The intercept gives a check on M, and the slope gives R_G.

In Table 9.1 some values of M and R_G determined by light scattering are given. Although R_G itself is unambiguous, its interpretation in terms of dimen-

TABLE 9.1   *Representative data from light scatteirng and low-angle X-ray scattering[a]*

Material	M_w	R_G (Å)
Lysozyme	14,100	15.2
β-Lactoglobulin	36,000	
	36,700	21.7
Serum albumin	70,000	29.8
Myosin	493,000	468
DNA sample	4×10^6	1170
Tobacco mosaic virus	39×10^6	924
Turnip yellow mosaic virus		104

[a]Underlined values are from low-angle X-ray scattering.

sions of the particles depends on particle shape. Table 9.2 lists some values of R_G in terms of particle dimensions. It should also be emphasized that the angular dependence of scattering will depend explicitly on shape if particles are so large that higher terms in the expansion in (9.43) have to be retained.

TABLE 9.2   *Relation between R_G and dimensions for particles of various shapes*

Shape	R_G	Where:
Sphere	$\sqrt{3/5}R$	R = radius of sphere
Prolate ellipsoid	$(\sqrt{2 + \gamma^{-2}/5})a$	Axes are $2a$, $2a$, $\gamma 2a$
Very long rod	$L/\sqrt{12}$	L = length of rod
Random coil	$(\overline{h^2})^{1/2}/\sqrt{6}$	$\overline{h^2}$ = mean-square end to end distance

In summary, light scattering provides a powerful tool for the determination of particle weight and dimensions. Its use is limited at one extreme by the low scattering power of small molecules and at the other by the difficulties encountered in interpretation for very large macromolecules. If the particles are very large, Equation (9.43) no longer adequately describes the angular dependence, which then reflects the particle shape as well. If the shape is not known, or if experimental difficulties prevent measurement to θ low enough for Equation (9.43) to be valid, a clear extrapolation to $\theta = 0$ may not be possible. In practice, it is very difficult, because of diffraction effects, to measure scattering at angles less than about 5°.

9.5 Low-Angle X-ray Scattering

While the *weights* of small molecules can be determined from the scattering of visible light, the method fails if we seek to use the angular dependence to measure the *dimensions* of particles with R_G less than about 100 Å. Since the resolving power of the method depends on the ratio $(R_G/\lambda)^2$, an obvious solution would be to use light of shorter wavelengths. But here nature has been uncooperative, for those materials that we would most like to investigate (proteins and nucleic acids, for example) begin to absorb very strongly a little below 300 nm. Below about 160 nm, almost everything, including water and air, absorbs very strongly, leaving a considerable region of the spectrum at the present time inaccessible to us. It is not until we reach the X-ray region that materials again become generally transparent. But here the scattering situation is very different; the interference effects, which were second order in the visible, now dominate the scattering. For example, the wavelength of the Cu-α radiation (a commonly used X radiation) is 1.54 Å, so that the distance across a ribonuclease molecule (molecular weight of 13,683) is of the order of 10λ. Clearly, an approximation such as Equation (9.43) will fail even at quite small angles, and the complete Equation (9.39) must be used. The situation is best illustrated by a specific example. Evaluation of the quantity $\sum \sum \sin hR_{ij}/hR_{ij}$ for atoms distributed uniformly in a solid sphere of radius R leads to the $P(\theta)$ function shown in Figure 9.9. For this example, we have chosen $\lambda = 1.54$ Å and $R = 15.4$ Å. The scatter-

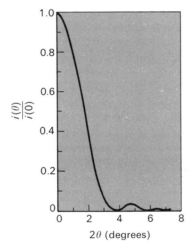

Figure 9.9 A graph of the scattering intensity versus angle for the scattering of 1.54-Å X rays from spheres of radius 15.4 Å.

ing is almost entirely confined to a very narrow angular range. The effect of the interference of waves is emphasized by the existence of maxima and minima in the $P(\theta)$ function.

Fortunately, it is possible to collimate an X-ray beam very well, so measurements can often be made to angles low enough to allow a decent extrapolation to $\theta = 0$. If this is done, the molecular weight can be calculated from the zero angle scattering, much as in light-scattering experiments. The theory looks a little different because it is most convenient to consider the individual electrons in the molecule as scatterers. By classical theory the intensity of X-ray scattering from a single electron is

$$i_e(\theta) = \frac{I_0}{r^2}\left(\frac{e^2}{mc^2}\right)^2 \frac{1 + \cos^2 \theta}{2} \tag{9.49}$$

where e is the electron charge, m the electron mass and c is the velocity of light. For a molecule containing n electrons, we shall have an intensity n^2 times as great, for the scattering is coherent and *amplitudes* must be added:

$$i_m(0) = \frac{I_0}{r^2}\left(\frac{e^2}{mc^2}\right)^2 n^2 \qquad (9.50)$$

where θ has been taken† as $0°$. To calculate the scattering from 1 cm³ of a solution containing N molecules per cubic centimeter, we add the *intensities* (assuming the molecules to be independent of one another):

$$i(0) = \frac{I_0}{r^2}\left(\frac{e^2}{mc^2}\right)^2 n^2 N \qquad (9.51)$$

We are interested in solutions, so we replace n by $n - n_0$, the excess electrons in a solute molecule over the volume of solvent it displaces. Also, we divide and multiply by M^2:

$$i(0) = \frac{I_0}{r^2}\left(\frac{e^2}{mc^2}\right)^2 NM^2\left(\frac{n - n_0}{M}\right)^2 \qquad (9.52)$$

Or, recalling that $N = \mathfrak{N}C/M$, where C is the weight concentration and $\mathfrak{N}$ is Avogadro's number,

$$i(0) = \frac{I_0}{r^2}\left(\frac{e^2}{mc^2}\right)^2 \mathfrak{N}\left(\frac{n - n_0}{M}\right)^2 MC \qquad (9.53)$$

The similarity to the light-scattering equation is obvious. The quantity $(n - n_0/M)$ depends on the chemical composition, since it is essentially a ratio of atomic numbers to atomic weights. As in the case of light scattering, it is necessary to extrapolate data to $C = 0$ because of nonideality of real solutions.

The angular dependence of the scattering in the very low-angle range will give the radius of gyration. The extrapolation might be made by graphing $i(\theta)$ versus $\sin^2(\theta/2)$ as in light-scattering experiments, but it has become customary to attempt to extend the range of a linear graph by a slightly different method. The first two terms of Equation (9.44) are identical with those in an expansion of $e^{-h^2 R_G^2/3}$; that is,

$$e^{-h^2 R_G^2/3} = 1 - \frac{h^2 R_G^2}{3} + \frac{h^4 R_G^4}{18} - \cdots \qquad (9.54)$$

For spherical particles, even the third term is identical to the expansion of the exponential. Thus to a good approximation for any particle in the low-angle range, we can write

$$P(\theta) \simeq e^{-h^2 R_G^2/3} \qquad (9.55)$$

or

$$\ln P(\theta) = -\frac{h^2 R_G^2}{3} = -\frac{16\pi^4 R_G^2 \sin^2(\theta/2)}{3\lambda^2} \qquad (9.56)$$

†By concentrating on the zero angle scattering here, we eliminate the problems of phase differences between scattering points.

wihch means that graphing $\ln i(\theta)$ versus $\sin^2 (\theta/2)$ will give a straight line with intercept $\ln i(0)$ and slope $-16\pi^4 R_G^2/3\lambda^2$. Such a graph is known as a *Guinier plot*. In Table 9.1 some molecular weights and particle dimensions obtained by low-angle X-ray scattering are given. Even the smallest protein molecules can be investigated by this method.

The angular dependence of the scattering at somewhat larger angles becomes dependent on the shape of the particles. While the analysis has not been carried as yet to the point where all of the potentially available information can be obtained, it is possible to determine such details as the cross-sectional area and mass per unit length of elongated particles.

9.6 Neutron Scattering

According to the basic principles of quantum mechanics, every "particle" has also a wavelike character. The de Broglie equation states that a particle of mass *m*, moving with velocity *v* can be considered to have an associated wavelength

$$\lambda = \frac{h}{mv} \tag{9.57}$$

where *h* is Planck's constant. Nuclear reactors can produce beams of neutrons with quite well-defined velocities, corresponding to wavelengths of the order of 2 to 4 Å. Such a "monochromatic" neutron beam can be used for scattering experiments. Since a nuclear reactor is much less readily available to most biochemists than a conventional light source, laser, or even X-ray source, it may seem surprising that neutron scattering has become an important tool in biophysical research. The reason lies in some very special features of the scattering of neutrons by matter. To give even a qualitative explanation, it is necessary to introduce a bit of nomenclature utilized in describing neutron scattering.

In Equation (9.49), which gives the scattering by an electron, the intensity is proportional to the square of the quantity e^2/mc^2. The *amplitude* of the scattered X ray is thus proportional to this number, which has the dimensions of length. A similar *scattering length* is hidden in Equation (9.5), as α/λ^2 (polarizability always has the dimensions of volume). The scattering lengths (b_x) that various atoms have for X rays are given approximately by

$$b_x = \frac{e^2}{mc^2}n \tag{9.58}$$

where *n* is the number of electrons in the atom, that is, the atomic number. There are two consequences of this relationship for X-ray scattering: (1) Since hydrogen has a very small value for b_x, hydrogen atoms do not contribute greatly to the scattering, and (2) the other common elements in most biopolymers (C, N, O) are roughly the same in scattering power (see Table 9.3).

TABLE 9.3   *X-ray and neutron-scattering lengths*[a]

Element	b_x	b_n
H	3.8	−3.74
D	2.8	6.67
C	16.9	6.65
N	19.7	9.40
O	22.5	5.80
P	42.3	5.10

[a]All lengths in units of 10^{-13} cm.

The point to this discussion is that the situation is very different for neutrons. Table 9.3 lists neutron scattering lengths for the common elements in biopolymers. The most striking observation is the negative value for hydrogen. Physically, this means that the neutron scattering from a hydrogen atom will be 180° out of phase with that from the other atoms listed. Since, as in X-ray scattering, we must add the amplitudes of scattered waves from different atoms in the molecule, the presence of hydrogen atoms has a major effect. In particular, it means that the scattering from H_2O is very different from that from D_2O.

Recall that it is always the *contrast* between solvent and macromolecule that allows us to observe the scattering. For example, if we change the polarizability of the solvent in a light-scattering experiment (by adding sucrose, for example) the scattering intensity will change. However, the effects that can be produced in this case are small. With neutron scattering, the situation is very different. The great difference in H_2O and D_2O scattering allows "matching" of the solvent background to the scattering from a macromolecule. In effect, the macromolecule can be made to "disappear" by choice of a suitable H_2O–D_2O mixture for solvent.

This would be only an interesting curiosity were there not another aspect. As Table 9.4 shows, the scattering length densities of proteins, nucleic acids, lipids, and carbohydrates all differ. This can be of enormous utility in studying complex macromolecules and particles. It is possible, using H_2O–D_2O mixtures of different compositions, to produce matching selectively with each type of material listed in Table 9.4. Consider, for example, the nucleosome depicted in Figure 9.10. Such a particle contains both protein and nucleic acid, but the relative disposition of these two components was originally unknown. Neutron-

TABLE 9.4   *Neutron-scattering lengths per unit volume*

Substance	$b/v \times 10^{14}$ $(cm^3/\mathring{A}^3)$
H_2O	−0.6
D_2O	6.4
Fatty acid	0.0
Protein	3.1
Carbohydrate	4.3
Nucleic acid	4.4

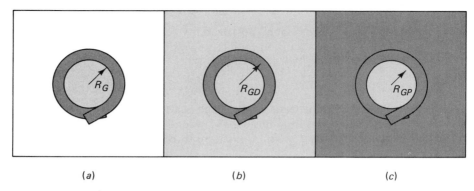

Figure 9.10 The principle of contrast matching in neutron scattering. A nucleosome, which has DNA (dark) wrapped about a protein core (lighter), is depicted schematically. (*a*) Solvent of low scattering power; the entire particle contrasts, and R_G is for the whole structure. (*b*) Solvent matched to protein; the R_{GD} measured is that of the DNA. (*c*) Solvent matched for DNA; the DNA "disappears" and the R_{GP} of the protein is measured.

scattering experiments were performed in D_2O–H_2O mixtures and Guinier plots used to obtain the radius of gyration. Under conditions where the mixture matched the scattering length density of nucleic acid, only the protein contributed to the scattering. Thus the R_G of the protein portion was obtained. On the other hand, under conditions where the protein was matched, the R_G of the nucleic acid portion could be found. The results (Figure 9.10) show unequivocally that the nucleic acid lies on the outside of the particle.

Similar experiments with lipoproteins have been used to demonstrate the separation between lipid and protein in such structures. Other variants of the experiment are possible. For example, if selected subunits in a multi-subunit complex are deuterated, it is possible to focus attention on these particular parts of a complicated structure. The potentialities of neutron scattering are obviously great. The main problem at the present time is that there are only a few reactors in the world which provide sufficient neutron flux to allow such experiments to be performed on a practicable time scale.

9.7 Raman Scattering

In the preceding sections, we have made the implicit assumption that the scattered radiation is at the same wavelength and frequency as the incident light. However, this need not always be true. In writing Equation (9.4) it was assumed that the polarizability, α, was a constant. However, closer examination shows that in many vibrating molecules, the polarizability depends to a small extent on the internuclear distance. Consider, for example, a diatomic molecule or group undergoing vibrational deformation with a frequency v'. If the polarizability depends on the bond deformation, we should write

$$\alpha(v') = \alpha + \beta \cos 2\pi v' t \tag{9.59}$$

where α is the "static" polarizability. If we then rewrite Equation (9.4), we find for the dipole induced by incident radiation

$$\boldsymbol{\mu} = \alpha(v')\mathbf{E}$$

$$= \alpha \mathbf{E}_0 \cos 2\pi v t + \beta \mathbf{E}_0 \cos 2\pi v t \cos 2\pi v' t \tag{9.60}$$

The first term leads to Rayleigh scattering, at the same frequency, v, as the incident light. But consider the second term: Since $\cos x \cos y = \frac{1}{2}[\cos(x + y) + \cos(x - y)]$, we find this term to be

$$\frac{\beta \mathbf{E}_0}{2}[\cos 2\pi(v + v')t + \cos 2\pi(v - v')t] \tag{9.61}$$

This suggests that the Rayleigh scattering at frequency v will be accompanied by scattering at frequencies $(v + v')$ and $(v - v')$. Since v' is a vibration frequency of the molecule, we may expect to observe in the spectrum of the scattered light a series of bands corresponding to the various vibrational frequencies which the molecule exhibits. Such bands constitute the *Raman spectrum* of the molecule. They will occur in two series, one at higher frequencies than the exciting light $(v + v')$, and the other at lower frequencies $(v - v')$.

The Raman bands are much less intense than the Rayleigh band, since $\beta \ll \alpha$ (by a factor of 10^{-4} or smaller). However, they are important, for they provide an adjunct to infrared (IR) spectroscopy. In fact, bands can be observed in the Raman spectrum that cannot be found in the IR spectrum. This is because the only requirement for Raman excitation is that the polarizability change upon bond deformation. For a vibration to be active in IR absorption, on the other hand, requires that a displacement of the center of charge occur during the vibration; there must be a finite transition moment. As an example, consider the symmetrical stretching vibration of the simple molecule O_2. Since the molecule has a center of symmetry, no displacement of the charge center occurs. This vibration is IR inactive, and no absorption band can be observed. However, the polarizability of O_2 changes as the $O{=}O$ bond is stretched, so Raman bands can be observed.

There is another advantage to Raman spectroscopy in the study of biologically important molecules. It will be recalled from Chapter 8 that IR spectroscopy is rendered difficult by the strong absorption of water in spectral regions of considerable interest. In contrast, the Raman bands of water are very weak, largely eliminating the interference and allowing studies in aqueous solution.

Until recently, Raman spectroscopy of macromolecules was almost impractical because of the intense Rayleigh scattering from large molecules. Using conventional light sources, the breadth of the exciting band was such that many weak Raman bands were overwhelmed by the edges of the enormous Rayleigh band. However, with the development of intense laser sources, which are both well collimated and highly monochromatic, Raman spectroscopy of proteins and nucleic acids has become a practical and fruitful technique. An example of such an application is shown in Figure 9.11.

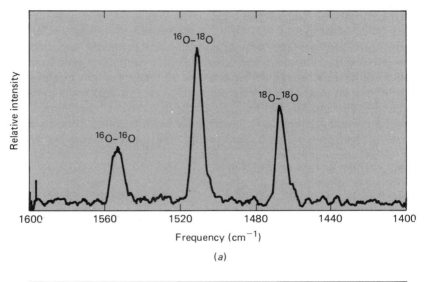

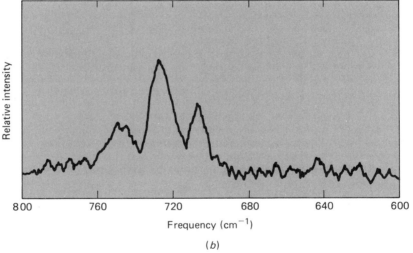

Figure 9.11 An example of an application of Raman spectroscopy to a biochemical problem. The mode of oxygen binding to a molluscan hemocyanin is being studied. (*a*) The Raman spectrum of gaseous O_2, in which a mixture of $^{16}O-^{16}O$, $^{16}O-^{18}O$, and $^{18}O-^{18}O$ molecules has been used. The three bands (from left to right) correspond to these three molecular species. (*b*) The resonance Raman spectrum of the oxyhemocyanin. The bands shown are absent in the deoxygenated protein. Note that the bands are shifted to much lower frequencies. This indicates that O_2 is bound as the peroxide ion, O_2^{2-}; corresponding bands in Na_2O_2 occur at about 740 cm^{-1}. The fact that only three bands are observed with oxyhemocyanin, with the expected spacings for isotope shifts, is evidence that the binding site is symmetrical; oxygen is bound between two nearly equivalent copper atoms. After T. J. Thannann, J. S. Loehr, and T. M. Loehr, *J. Am. Chem. Soc.*, **99**, 4187–4189 (1977). Reprinted by permission of the author and the American Chemical Society.

The intensity of Raman bands can often be enhanced by taking advantage of the *resonance Raman effect*. In the neighborhood of an electronic absorption band of a chromophore, the polarizability will become very large, and the Raman scattering from vibrational modes of the chromophore will be intensified. This explains the intensity of the bands shown in Figure 9.11(*b*). Excitation was at a wavelength of 530.9 nm, which lies within an absorption band of the copper–oxygen chromophore of this protein. The resonance Raman effect allows one to single out particular chromophores in a complex molecule for selective study.

PROBLEMS

*1.   At one time, it was conventional to describe light scattering in terms of the *turbidity*, τ. This is defined in analogy to the extinction coefficient in spectroscopy; if a beam of incident intensity I_0 passes through 1 cm³ of solution and emerges with a smaller intensity I (the loss being by scattering), we define

$$\tau = -\ln \frac{I}{I_0}$$

Show that for Rayleigh scattering, $\tau = (16\pi/3)R_\theta$. [*Hint:* Integration of the scattered intensity over a sphere at distance r from the solution should give the total loss of energy but this should also be equal to $I - I_0$ as defined above. Why?)]

2.   A complication in the absorption spectroscopy of very large molecules arises from the light lost by scattering. Calculate the absorbance due to *scattering* alone, for a solution of a hemocyanin, given the following data:
   1. Cell thickness in direction of beam = 1 cm.
   2. Protein concentration = 0.01 g/ml, M = 3,800,000.
   3. n_0 = 1.33, dn/dC = 0.198 ml/g, λ = 2800 Å.
   You may assume Rayleigh scattering. See Problem 1.

3.   To visualize the effect of dust contamination on light-scattering experiments, perform the following calculation. A solution, containing 5 mg/ml of protein of M = 100,000, is contaminated to the extent of 0.001 percent of the protein weight by dust. The dust particles are 0.1 μm in radius, with density = 2.00. By what percent will the 90° scattering be changed by this contamination, assuming Rayleigh scattering for all particles?

*4.   To provide a specific example to illustrate the general problem of large-particle scattering, let us consider the scattering from a dumbbell-shaped molecule containing

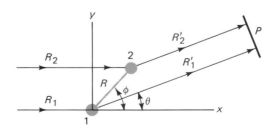

two scattering centers separated by a distance R. To simplify the geometry, we shall require that the molecule remain in the plane of the paper but assume that it is in rapid Brownian rotation in that plane. The molecule assumes an angle ϕ to the x axis, with one of its scattering centers at the center of the coordinate system. We consider incident light parallel to the x axis, scattered at an angle θ to this axis. Calculate the function $P(\theta)$ for this model, making any necessary approximations. [*Hint:* The scattering angle θ can vary from zero to π, and we must consider all orientations of the particle with respect to the x axis. The path difference at P for R_2' and R_1' is best expressed in terms of an angle ψ, such that $\phi = \psi + \theta/2$.]

5. The radius of gyration for a circle is just the radius of the circle. The expression for the R_G of a rod is given in the text. Suppose that a DNA molecule could exist either as a rigid rod 1000 Å long or as a circle with this circumference. Calculate the ratio of scattering at $\theta = 90°$ for the rod form to that for the circle form. Assume that $\lambda_0 = 5460$ Å. If a 1 percent difference is detectable, would the light-scattering measurement serve to detect the breaking of circles?

6. The following data have been given [P. Outer, C. T. Carr, and B. H. Zimm, *J. Chem. Phys.*, **18**, 830 (1950)] for the light scattering of polystyrene in butanone at 20°C. The light was polarized perpendicular to the plane of measurement, that is, along the z axis. The following data are given for various angles and concentrations:

Values of $CI_0/r^2 i_\theta$

	C (g/ml)			
θ (deg)	0.312×10^{-3}	0.624×10^{-3}	1.25×10^{-3}	2.50×10^{-3}
26	1.58	1.65	1.86	2.21
36.9	1.66	1.76	1.98	2.33
66.4	1.82	1.90	2.07	2.52
90	1.95	2.08	2.24	2.69
113.6	2.17	2.24	2.47	2.88

$n_0 = 1.378$, $dn/dC = 0.214$ ml/g, and $\lambda = 5460$ Å. Obtain R_G and $\overline{M}_w$. Be sure to use correct equations for polarized light. $P(\theta)$ does *not* depend on polarization.

7. The 50S ribosomal subunits from *E. coli* have been studied by low-angle X-ray scattering. X rays of $\lambda = 1.54$ Å were used, and the following data recorded:

θ (mrad)	I/I_0
1.0	0.98
1.5	0.93
2.0	0.86
2.5	0.81
3.0	0.74
3.5	0.65

Calculate the radius of gyration. If the particles are spheres, what is their diameter?

REFERENCES

Scattering Processes: General

Cantor, C. R. and P. R. Schimmel: *Biophysical Chemistry*, W. H. Freeman and Company, Publishers, San Francisco, 1980, Chaps. 13 and 14. Rigorous and comprehensive.

Feynman, R. P., R. B. Leighton, and N. Sands: *The Feynman Lectures on Physics*, Addison-Wesley Publishing Company, Inc., Reading, Mass., 1963, Chaps. 28–32. A remarkably lucid discussion of the fundamentals. The entire set of three volumes is strongly recommended.

Kauzmann, W.: *Quantum Chemistry*, Academic Press, Inc., New York, 1957, Chap. 15. A very clear treatment.

Light Scattering

Pittz, E. P., J. C. Lee, B. Bablouzian, R. Townsend, and S. N. Timasheff: "Light Scattering and Differential Refractometry," in *Advances in Enzymology* (C. H. W. Hirs and S. N. Timasheff, eds.) Academic Press, Inc., New York, 1973, Vol. 27, pp. 209–256.

Schurr, J. M.: "Dynamic Light Scattering of Biopolymers and Biocolloids," in *CRC Crit. Rev. Biochem.* (G. Fasman, ed.), **4**, 371–431 (1977).

Stacey, K. A.: *Light Scattering in Physical Chemistry*, Academic Press, Inc., New York, 1956. Still a useful introduction, although somewhat dated insofar as techniques are concerned.

Tanford, C.: *Physical Chemistry of Macromolecules*, John Wiley & Sons, Inc., New York, 1961, Chap. 5. Detailed and careful derivations of the fundamental equations. The analysis of Rayleigh scattering given in this book draws heavily on Tanford's treatment.

Zimm, B. H.: *J. Chem. Phys.*, **16**, 1093–1099 (1948). Development of Zimm's methods for analysis of angular dependence.

Other Scattering Methods

Pessen, H., T. F. Kumonsinski, and S. N. Timasheff: "Small Angle X-Ray Scattering," in *Methods in Enzymology* (C. H. W. Hirs and S. N. Timasheff, eds.), Academic Press, Inc., New York, 1973, Vol. 27, pp. 151–209.

Pilz, I., O. Glattner, and O. Kratky: "Small Angle X-Ray Scattering," in *Methods in Enzymology* (C. H. W. Hirs and S. N. Timasheff, eds.), Academic Press, Inc., New York, 1979, Vol. 61, pp. 148–249. Together with the reference above, these give a quite complete coverage of current methods in low-angle scattering.

Schoenborn, B. P. and A. C. Nunes: "Neutron Scattering," *Annu. Rev. Biophys. Bioeng.*, **1**, 529–549 (1972).

Van Wart, H. E. and H. A. Scheraga: "Raman and Resonance Raman Spectroscopy," in *Methods in Enzymology* (C. H. W. Hirs and S. N. Timasheff, eds.), Academic Press, Inc., New York, 1978, Vol. 49, pp. 67–149.

Chapter Ten

CIRCULAR DICHROISM

AND OPTICAL ROTATION

One feature shared by a wide variety of biopolymers is the existence of helical secondary structure. The α helices of polypeptides and proteins, the single, double, and triple helices of polynucleotides, and the helical conformations of some polysaccharides are both unique in each class of compounds and important to their biological function. The detection of such structure in biopolymers and quantitative measurement of the amount and changes in amount of such structures play an important role in our understanding of the way in which these molecules function.

Unfortuately, none of the physical properties we have discussed so far are very useful for the detection of helical structure. It is true that absorption spectra are often sensitive to regular secondary structure, via hypochromism and exciton splitting, but it would be difficult to rest the case for helical conformations on this kind of evidence alone. X-ray diffraction (Chapter 11) can yield detailed information on structure but is applicable only to the solid state. What is needed is a physical property that depends explicitly on the asymmetry of structure of dissolved molecules. This is provided by the response of molecules to polarized light.

10.1 Polarization of Radiation

An electromagnetic wave is characterized by the amplitude and orientation of its component electric and magnetic fields. As shown in Figure 10.1, the electric and magnetic field vectors will always be perpendicular to one another and perpendicular to the direction of propagation. It is by their orientation with respect to

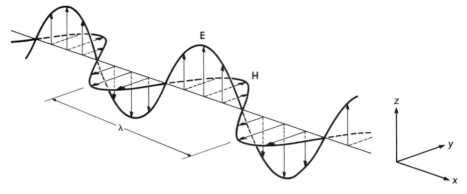

Figure 10.1 Plane-polarized radiation.

an external frame of reference (the apparatus, a molecule) that we describe the polarization. The simplest case is that of plane polarization (Figure 10.1); here the direction of propagation is along the x axis, the electric vectors point in the z direction, and the magnetic vectors point in the y direction. At a given point on the x axis, the field vectors oscillate with a frequency ν (reciprocal seconds).

Plane-polarized light is so easy to produce and work with that the significance of other kinds of polarization is often neglected. Equally important is circularly polarized light. In Figure 10.2 we show the electric field only. At any

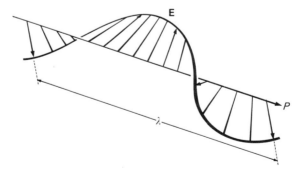

Figure 10.2 Circularly polarized radiation. Only the electric field is shown. By the convention used, this is right circular polarization, since an observer at point P would see the field rotating in a clockwise direction as the waves passed.

point on the x axis the electric vector can be visualized as rotating in a plane perpendicular to that axis, making v rotations per second. The retardation of successive vectors along the x axis is such that the wavelength λ corresponds to one complete rotation (note in Figure 10.2). The motion of a point of a field vector is a helical path, with a velocity of propagation in the x direction of $c = \lambda v$. Circular polarization can be either right- or left-handed, depending on the sense of rotation with respect to the direction of propagation. By convention, we consider light to be right circularly polarized when the electric vector, as seen looking toward the source, rotates in a clockwise direction.

If we combine left and right circularly polarized waves of equal amplitude, the resultant of the vector additions of the fields will be plane-polarized light (see Figure 10.3). This means that a plane-polarized beam can be considered the sum of two circularly polarized components. We shall use this fact later.

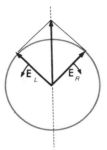

Figure 10.3 Plane-polarized light as the sum of two equal circularly polarized components. We are looking directly toward the source. The vector sum of the two field components will oscillate along the dotted line as $\mathbf{E}_L$ and $\mathbf{E}_R$ rotate.

If two circularly polarized components of unequal amplitude are combined (Figure 10.4), the result is *elliptically* polarized light. This will be characterized by both an ellipticity (determined by the ratio of axes of the ellipse) and the direction of the major axis. Obviously, in the extreme limits, elliptical polarization approaches either circular or plane polarization (see Section 10.2).

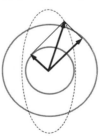

Figure 10.4 Elliptically polarized light.

It is equally easy to represent a circularly polarized wave as the sum of two plane-polarized waves, 90° out of phase and with their electric vectors perpendicular (see Figure 10.5).

Finally, *unpolarized* light can be considered as a wave in which there is no fixed direction of polarization. An incoherent source, which produces a melange

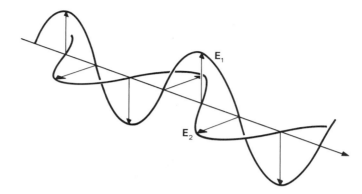

Figure 10.5 The same circularly polarized wave shown in Figure 10.2 is here represented as the sum of two plane-polarized waves. Only the electric fields are shown. Note how the electric vector, which will be the sum of $\mathbf{E}_1$ and $\mathbf{E}_2$, describes the same helical path as in Figure 10.2.

of waves of different polarization properties, will yield such radiation. Various optical devices may be used to separate from an unpolarized beam radiation with the several kinds of polarization characteristics described above. We turn now to the interaction of polarized light with molecules.

10.2   Circular Dichroism and Optical Rotation: Definitions

Some, but not all substances exhibit what is called *optical activity*. This is manifested in two seemingly different but closely related phenomena: circular dichroism and optical rotation.

CIRCULAR DICHROISM

If we pass circularly polarized light through a solution of an optically active substance, we shall find that the absorptivity depends on the handedness of the light. Some absorption bands may absorb more strongly the left circularly polarized beam; others may absorb more strongly the right. We define the circular dichroism (CD) at a given wavelength λ as $\Delta\epsilon = \epsilon_L - \epsilon_R$, the difference in extinction coefficients. A particular absorption band can be characterized by its *rotational strength*, which is essentially an integrated extinction coefficient difference over the band:

$$R = \frac{(2.303)(3000)hc}{32\pi^3 \mathfrak{N}} \int \frac{\Delta\epsilon}{\lambda} \, d\lambda \tag{10.1}$$

where h is Planck's constant and c is the velocity of light. It is important to note that circular dichroism is observed only in the wavelength regions where the substance absorbs light. However, unlike absorption bands, CD bands can be either positive or negative in sign (see Figure 10.6).

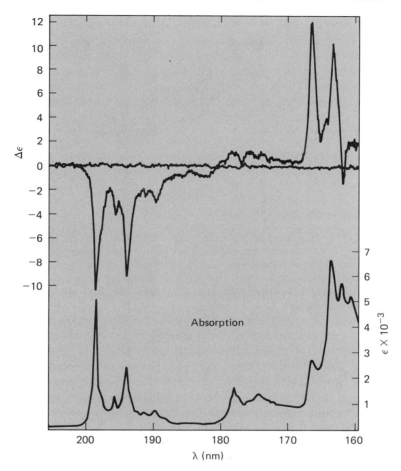

Figure 10.6 Absorption spectrum and CD spectrum of (+)-3-methylcyclopentane. The figure is included to emphasize three points: (1) CD bands can be of either positive or negative sign; (2) $\Delta\epsilon$ is several orders of magnitude smaller than ϵ; and (3) some bands which exhibit intense absorption are weak in CD, and vice versa; compare the band at about 178 nm to that at about 194 nm. After W. C. Johnson, *Rev. Sci. Instrum.*, **42**, 1283–1286 (1971). Reprinted by permission of the author and the American Institute of Physics.

OPTICAL ROTATION

If plane-polarized radiation of *any* wavelength is passed through an optically active substance, the plane of polarization will be rotated (see Figure 10.7). The angle of rotation (α) is found to depend on the nature of the substance, the thickness of the sample (d), and the concentration (C) of the optically active substances in the sample. Thus

$$\alpha = [\alpha]\, dC \tag{10.2}$$

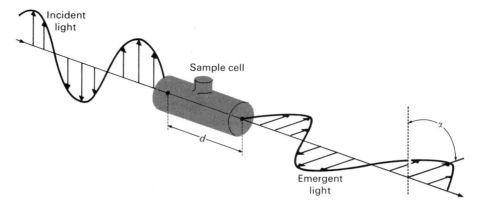

Figure 10.7 The phenomenon of optical rotation. The incident beam is plane-polarized, with electric vector in the *z* direction. The emergent light is still plane-polarized, but the plane has been rotated through an angle α. The solution is dextrorotatory.

Here *C* is measured in grams per milliliter and *d* (by old convention) in decimeters.

The quantity [α] so defined is called the *specific rotation*. Ideally, it depends only on the nature of the optically active substance. Note that rotation of the plane of polarization can be either clockwise or anticlockwise as one looks toward the source of radiation. The first is termed *dextrorotation* and is given a positive sign; the second is termed *levorotation* and is given a negative sign.

The relationship between circular dichroism and optical rotation is analogous to the relationship between the absorption and refraction of unpolarized light. It will be recalled, from Chapters 8 and 9, that the interaction of an electromagnetic wave with the electrons in a chromophore can be of two kinds: Far from an absorption band (resonance frequency) there are only induced oscillations, with accompanying scattering, and the forward scattering gives rise to the effect of refraction. Near a resonance frequency (absorption band) there is as well a loss of energy to the molecule; in the terminology of quantum mechanics, we say that there is a finite probability of excitation to a higher-energy state of the molecule. Now it should be clear that circular dichroism is an absorptive phenomenon. It remains to show that optical rotation is the consequence of refractive effects. If an optically active substance shows a difference in absorption between left and right circularly polarized light, it should be expected that there will be, at all wavelengths, a difference in refractive index for the two components. At the beginning of this chapter we saw that plane-polarized light could be represented as the sum of two circularly polarized beams, of equal amplitude and opposite sense (Figure 10.3). Consider what happens when a plane-polarized ray passes through an optically active substance: The left and right circularly polarized components will each be retarded by refraction, but one will be retarded more than the other. The effect is shown, in Figure 10.8, to be a rotation of the plane of polarization. Thus we may state the relationship between circular dichroism and optical rotation succinctly: Circular dichroism results from a difference in absorption of L and R. Optical rotation results from a difference in refractive index for L and R.

Within an absorption band, the situation becomes more complex. Now there are differences in both absorption *and* refractive index for left and right circularly polarized light. Again imagining a plane-polarized incident beam as made up of left and right circularly polarized components, we can see that one component will be both absorbed more, and retarded more than the other. The consequence, as shown in Figure 10.4, is that the light will become *elliptically polarized*; the point of the electric vector now describes an elliptical path. The major axis of the ellipse will be tilted from the original plane of polarization by an angle equal to the optical rotation at this wavelength. The eccentricity of the ellipse will obviously depend upon the difference in absorbance for L and R, and hence can be used as a measure of the circular dichroism. In the extreme case where one component is wholly absorbed, the plane-polarized light is transformed into circular polarization; in the opposite extreme, where there is refractive retardation but no absorption, the ellipse becomes infinitely thin, which is a way of describing the situation shown in Fig. 10.8. The *ellipticity*, θ,

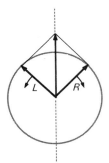

Incident light

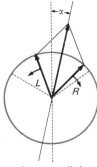

Emergent light

Figure 10.8 Optical rotation as the consequence of a different refractive index for left and right circularly polarized components. The situation is the same as in Figure 10.7. The substance has a higher refractive index for L than R; therefore the L component is retarded more and has been set back with respect to R. The sum of the components is still plane-polarized, but the plane has been rotated.

of elliptically polarized light is defined as the arctangent of the ratio of the minor to major axes of the ellipse. The *molar ellipticity*, $[\theta]$, is directly related to $\epsilon_L - \epsilon_R$:

$$[\theta] = 3300 \, \Delta\epsilon \tag{10.3}$$

where $\Delta\epsilon$ has the units liters/mole-cm and $[\theta]$ the units deg-cm²/decimole. One finds circular dichroism data expressed by some workers in terms of $\Delta\epsilon$; others customarily use $[\theta]$. In any event, it should be noted that circular dichroism is a small quantity, $\Delta\epsilon/\epsilon$ is usually of the order of 10^{-3} or lower (see Figure 10.6), and the ellipse of polarization is *very* thin. It is for this reason that the use of circular dichroism measurements to characterize optical activity is relatively new; sensitive instruments are required. Since $\epsilon_L - \epsilon_R = (A_L - A_R)/lc$, it is necessary to measure an absorbance difference of the order of $10^{-3} \times$ the total absorbance of the sample. Because of these instrumentation limitations, for a very long time optical activity was characterized in terms of optical rotation. This is rather like attempting to do absorption spectroscopy with a refractometer. The existence of optical rotation in the visible region of the spectrum points to the existence of optically active absorption bands at lower wavelengths, but tells nothing about their detailed structure.

10.3 Molecular Basis of Rotatory Power

Why do some substances exhibit different absorption and refractivity for left and right circular polarization? A clue can be found by considering the two molecules shown in Figure 10.9. These compounds are very similar in structure.

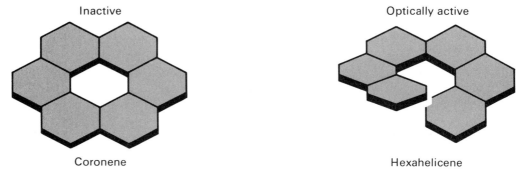

Coronene   Hexahelicene

Figure 10.9 Two similar substances, one of which is optically active, and the other inactive.

Both have absorption bands in the near ultraviolet, involving electron displacements in the conjugated ring systems. Yet coronene is optically inactive, whereas hexahelicene exhibits very large circular dichroism and optical rotation. Why is this?

One important and obvious difference between these molecules is that coronene is planar and has a plane of symmetry, whereas hexahelicene is asymmetric—in fact, it is helical. Let us consider what happens when an electron is driven in a helical path. In Figure 10.10 is shown an idealized molecule, which could be thought of as a prototype for any right-handed helical structure.

The helical molecule is pictured as interacting with right circularly polarized light. For the purpose of this discussion, we have chosen to represent circularly polarized light as the sum of two perpendicular plane-polarized components, differing in phase by 90°. Now consider the possible interaction of the radiation with the molecule. In the first place, the electric field of the radiation will induce an oscillating dipole on the helical path. Note that since the path is a helix, even an electric field parallel to the helix axis can induce such motion. (The electron goes up and down, as well as going around—there is a component of the induced electric dipole parallel to the helix axis.) But still another important consequence arises from the helical structure. An oscillating magnetic field parallel to the helix axis will induce a current in the helix. This will depend, according to the laws of electromagnetism, on the time derivative of the magnetic field. The magnetic field of circularly polarized light has a component parallel to the electric field (see the dashed curve in Figure 10.10). This is 90° out of phase with the electric

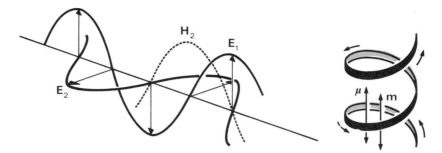

Figure 10.10 The interaction of a circularly polarized wave with a helical molecule. Electron paths are shown by arrows on the helix. The induced electric and magnetic moments parallel to the helix axis are denoted by μ and **m**. The magnetic field associated with the leading component of the radiation (H_2) is indicated by a dashed line.

field parallel to the helix axis. But its derivative will be *in* phase. Thus both the electric and magnetic fields of the radiation will contribute to an electron displacement on a helical path. In other words, electrons circulating about a helix create both an electric dipole moment and a magnetic dipole moment in the axis direction. But (and here is the crux of the matter) the phase relationship of the two plane-polarized components will differ by 180° for left and right circularly polarized light (at this point, see which phase the component E_1 would have to have in Figure 10.10 to yield *left* circularly polarized light). This means that whereas for one sense of polarization the electric and magnetic fields will be acting in concert on the electron, in the other sense of circular polarization they will be opposed. Thus we can qualitatively show that a molecule such as that in Figure 10.10 should exhibit different absorptivities and different refractive indices for left and right circularly polarized light. Of course, the argument is sketchy and incomplete. We should for example, consider different orientations of the molecule. But note that even when turned end for end, a right helix is still a right helix.

Another critical point emerges. If we had considered a helix of the opposite sense, the only difference would be that electron circulation would proceed in the opposite direction for a given electric field. This would reverse the magnetic dipole moment; therefore, a left-hand helix ought to display optical rotatory effects just the opposite to those given by a right-hand helix.†

†A molecule like hexahelicene can exist in either right- or left-hand helical forms, A *racemic* mixture, containing equal amounts of the two forms, would exhibit no optical activity.

The above argument is qualitative. For a more detailed treatment, the reader should consult Kauzmann (1957). But the result of a quantum-mechanical treatment can be stated very compactly. We have already encountered the concept of a transition dipole moment ($\mathbf{\mu}$), which we may associate with the charge displacement in a transition. If charge rotation occurs as well, an analogous quantity, the magnetic transition moment ($\mathbf{m}$), can be defined. The existence of optical rotatory power, from a quantum-mechanical point of view, depends on the magnitudes of these moments and their relative orientation. In fact, the rotational strength is given by

$$R = \mathrm{Im}\,(\mathbf{\mu} \cdot \mathbf{m}) \qquad (10.4)$$

where Im means the "imaginary part of" the dot product of the two vectors. (This nomenclature is necessary because the moments are complex vectors and their phase differs by 90°.) For purposes of calculation, we may say in most instances that R is given by $m\mu \cos\theta$, where m and μ are the vector magnitudes and θ is the angle between them. Thus, if the electric and magnetic moments have components in parallel, R is positive. If they are opposed, R is negative. If they are perpendicular, $R = 0$. This takes care of coronene very nicely, for the electron motion in such a planar molecule is in the plane, so the electric moment will lie in the plane and the magnetic moment will be perpendicular to it. In a simple physical sense the electric and magnetic fields are neither able to work together nor opposed upon the electron. Thus coronene is optically inactive.

Finally, it should be emphasized that it is a plane of symmetry that is critical for the absence of optical rotation. For even if substituents were placed above and below the plane of the coronene ring, with the ring plane as a mirror plane, and even if out-of-plane helical electron motions occurred in such substituents, these motions would be of opposite sense and thus cancel insofar as rotatory power was concerned. The *asymmetric carbon atom*, mentioned so often by carbohydrate chemists, now appears as a special case. If four different substituents are attached to a carbon, the static electric field they generate must be asymmetric. Then the path of any displaced electron must be a portion of some kind of a helix; therefore optical activity should result.

10.4 *Rotatory Behavior of Macromolecules*

It should be clear by now that optical activity (optical rotation, circular dichroism) is observed when and only when the environment in which a transition occurs is asymmetric. In macromolecules there are at least three kinds of asymmetry that can lead to optical activity:

1. The primary structure may be inherently asymmetric. For example, transitions involving electrons in the vicinity of the α carbons of amino acids and polypeptides will be optically active because of the inherent asymmetry at this point; the α carbons of most amino acids have four different substituents.

2. The secondary structures of many biopolymers are helical. This may result in optical activity for electronic transitions either in the chain backbone or in helically arrayed side groups.

3. The tertiary structure of a macromolecule may be such that an inherently symmetric group is thrust into an asymmetric environment. For example, transitions involving the ring electrons in tyrosine are normally only weakly optically active, but in some globular proteins the surroundings of a buried tyrosine are such as to produce asymmetrical electrical fields about the ring. These may distort the electron displacement in the transition so as to lead to strong optical activity.

It is the second example above, the influence of helical secondary structure on optical activity, that has received the greatest attention to date.

The behavior of helical structures has already been alluded to above. The electron-on-a-helix analog is useful in describing the behavior of some kinds of helical polymers. As another example of the effect of helical secondary structure on optical activity, we shall use a case similar to that considered in Chapter 8, a dimer in which there is interaction between transition dipoles (see Figure 10.11). This situation resembles that found in polynucleotides. Considering transitions in the planes of the rings, we recall from Chapter 8 that a single transition in the monomer will be split into two transitions in the dimer—a simple form of exciton splitting. These two modes of oscillation are shown in Figure 10.11. Now the

Figure 10.11 A schematic picture of a dimer in which electronic transitions in the two rings are coupled. The rings are parallel but rotated with respect to one another. In-phase and out-of-phase oscillations are shown.

transition moments are not coplanar, since the rings, while parallel, have been rotated with respect to one another (the beginning of helical structure). Thus in either mode of oscillation there will be a rotation of charge as well as a translation of charge.† This means that there will be a magnetic dipole as well as an electric dipole associated with each mode. But note also that while in one mode the electric and magnetic moments are in the same direction, in the other they are opposed. This will result in opposite rotational strengths in the two modes, with $R_1 = -R_2$. Therefore, the circular dichroic spectrum will resemble Figure 10.12.

For larger helices, involving larger numbers of residues, each rotated with respect to its neighbor, the splitting of the transitions becomes more complex.

---

†It will be worthwhile at this point to examine the directions of electron displacement and charge rotation to see how the two modes of oscillation differ.

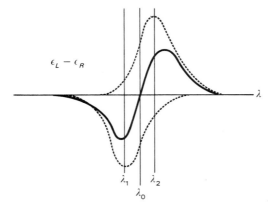

Figure 10.12 The kind of CD spectrum expected for a dimer exhibiting exciton splitting of a transition at λ_0 into two bands at λ_1 and λ_2. The individual CD bands are shown by the broken lines and the overall CD curve by the solid line.

However, the shape of the CD curves remains very much the same. In Figure 10.13 is shown the CD spectrum of polyriboadenylic acid, compared to that for the dimer and monomer. Note that the optical activity exhibited by the polymer is much greater and qualitatively different from that of the monomer.

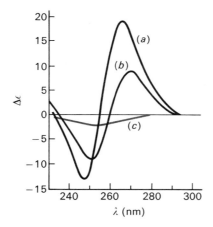

Figure 10.13 Circular dichroism of polyriboadenylic acid. Curve (a), high polymer; curve (b), dimer; curve (c), monomer. All are at neutral pH, with $\Delta\epsilon$ on a per-mole-of-residues basis. Data from K. E. van Holde, J. Brahms, and A. M. Michelson, *J. Mol. Biol.*, **12**, 726 (1965).

It can be shown that if the exciton transitions do not interact with other bands, then rotational behavior will be conservative; that is, we shall expect for the total group of components,

$$\sum_k R_k = 0 \qquad (10.5)$$

This situation, in which we have closely spaced series of bands of equal and opposite rotational strength, is approximated in many helical polymers. The circular dichroism of polynucleotides is very sensitivie to changes in helical conformation and has been widely used to monitor such changes in solution. An extreme example is shown in Figure 10.14. In dilute salt solution the poly-nucleotide poly (dG·dC)·poly(dG·dC) exists as a right-hand double helix, rather like that of B-form DNA. But in concentrated salt solutions, a transition

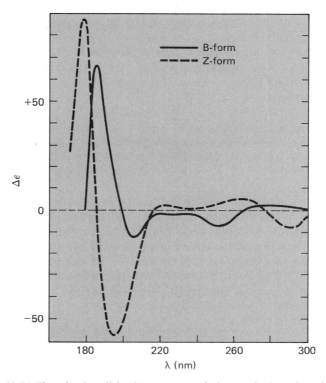

Figure 10.14 The circular dichroism spectra of the synthetic polynucleotide poly(dG·dC)·poly(dG·dC) in the B form (low ionic strength) and in the Z conformation adopted at high ionic strength. Data from I. Anderson, R. Searles, W. A. Baase, and W. C. Johnson.

occurs to a *left* helix, the so-called "Z form." This is reflected by a near inversion of the CD spectrum, as would be expected from our previous discussion of the circular dichroism of helical structures. The inversion is not exact, because Z-DNA is not just left-handed B-DNA; it has a somewhat different arrangement of the phosphodiester backbone, and altered base stacking. These differences are reflected in the different shape of the CD spectrum.

Although it is not yet possible to deduce *details* of secondary structure from CD spectra, it should be obvious from this example that circular dichroism can be a powerful tool for following such conformational changes in solution.

10.5 Determination of the Secondary Structures of Polypeptides and Proteins

All of the amino acid residues (with the sole exception of gylcine) are asymmetric units and hence optically active. In addition, organized secondary structures such as the α helix, β sheets, and β turns make up a substantial part of most native proteins and some synthetic polypeptides under appropriate conditions. Finally, side-chain chromophores, such as those in the aromatic amino acids, as well as prosthetic groups (such as the heme in heme proteins) may lie in asymmetric environments and hence exhibit optical activity. Thus it is not surprising that proteins and polypeptides exhibit complicated and distinctive CD spectra.

We shall center our attention on the polypeptide backbone itself, for in most cases the optical activity associated with this structure, in its various conformational states, dominates the CD spectrum. Three kinds of electronic transitions are recognized as being associated with the peptide group (Figure 10.15):

1. At about 210 to 220 nm, there is an $n \rightarrow \pi^*$ transition, in which nonbonding electrons of the carbonyl oxygen are promoted to an antibonding π orbital. This is a weak transition in absorbance, but it can play a major role in the CD spectrum (see Figure 10.15).

2. There is a very strong $\pi \rightarrow \pi^*$ transition centered at about 190 nm.

3. Below 180 nm, the structure of peptide spectra is less well understood. A transition (probably also $\pi \rightarrow \pi^*$) has been obverved at about 160 nm.

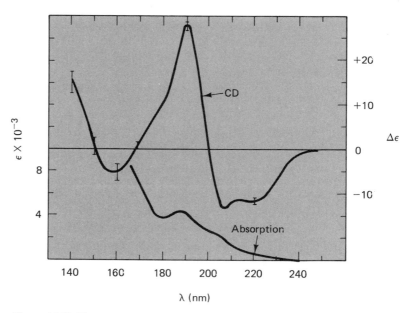

Figure 10.15 The vacuum ultraviolet absorption spectrum and CD spectrum of the synthetic polypeptide poly-γ-benzyl-L-glutamate in the α-helical conformation. Data from W. C. Johnson and I. Tinoco, *J. Am. Chem. Soc.*, **94**, 4389–4390 (1972). Reprinted by permission of the authors and the American Chemical Society.

However, both absorption spectroscopy and CD are difficult to measure in this range.

Depending on the secondary structure in which a particular peptide group is embedded, these transitions contribute in very different ways to the overall CD of the macromolecule. Figure 10.16 depicts CD spectra of the synthetic polypeptide, poly-L-lysine, in α-helical, β-sheet, and random-coil forms. The curves are remarkably different, and suggest that circular dichroism might be a powerful tool for analyzing the secondary structure in proteins. Many schemes have been proposed, based either on curves like those in Figure 10.16, obtained with synthetic polypeptides in "pure" conformations, or upon attempts to deduce spectra for the idealized structures from CD spectra of proteins whose secondary structure is known from X-ray crystallography. Such efforts have

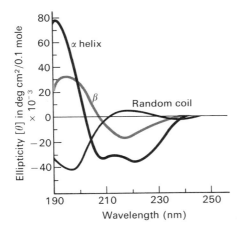

Figure 10.16 Circular dichroism of a polypeptide (poly-L-lysine) in various conformations. After N. Greenfield and G. D. Fasman, *Biochemistry*, **8**, 4108 (1969). Copyright © 1969 by the American Chemical Society.

been only moderately successful, for a number of reasons. First, some structural features, such as β turns, cannot be obtained in "pure" form in polypeptides—one cannot have a synthetic polypeptide that is *all* β-turn. Second, the CD spectrum observed for a particular kind of structure (that is, β sheet) seems to differ considerably from one model compound to another. Finally, there are almost certainly effects of helix length, environment in the protein, and so forth, which cannot be readily accommodated in such an analysis.

For such reasons as these, other and more empirical approaches have been used recently to relate circular dichroic spectra to protein secondary structure. For example, Hennessey and Johnson (1981) have used the data from X-ray diffraction studies on a number of proteins to deduce "basis spectra" which can be used to fit CD curves. None of these basis spectra corresponds to a single secondary structure, yet each can be regarded as a linear combination of such spectra. Since the relative amounts of α helix, β sheet, and so forth, are known for each protein of the basis set, the relative weightings of the basis spectra needed to fit the observed spectrum can reveal the secondary structure of the

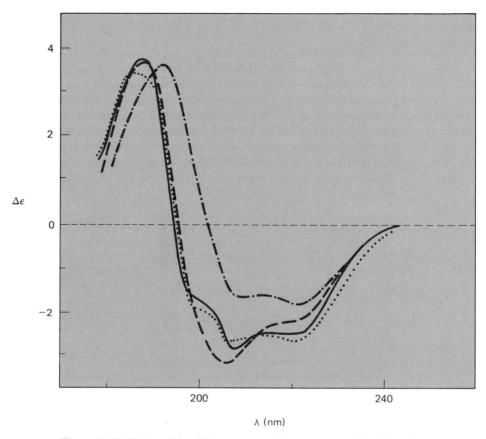

Figure 10.17 Fitting of the CD spectrum of the protein papain (—) using one (–·–·–), three (– – –), or five (·····) of the "basis" spectra of Hennessy and Johnson. Adapted from J. P. Hennessy and W. C. Johnson, *Biochemistry*, **20**, 1085–1094 (1981). Reprinted by permission of the authors and the American Chemical Society.

unknown protein. Figure 10.17 shows the fitting of the CD spectrum of the enzyme papain, obtained by using successively more of the basis set; the final fit is very good. Table 10.1 shows the agreement obtained between X-ray diffraction and CD analysis for a number of proteins. It should be emphasized that data to quite low wavelengths ($\sim$180 nm) are essential to provide enough information for an analysis of this kind. If such data are available, it seems that circular dichroism can provide quite detailed information about protein secondary structure. However, the technique is relatively unrevealing with respect to tertiary or higher structure. For such information, one must turn to X-ray diffraction, as described in Chapter 11.

TABLE 10.1 *Deduction of protein secondary structure from CD measurements[a]*

Protein	Method	H	A	P	I	II	III	T	O
						Conformation[b]			
Cytochrome *c*	X-ray	0.38	0.00	0.00	0.03	0.06	0.01	0.07	0.45
	CD	0.42	0.01	0.02	0.05	0.03	0.03	0.04	0.40
Elastase	X-ray	0.10	0.27	0.00	0.09	0.07	0.03	0.03	0.41
	CD	0.10	0.28	0.01	0.09	0.05	0.03	0.05	0.39
Hemoglobin	X-ray	0.75	0.00	0.00	0.08	0.01	0.04	0.01	0.11
	CD	0.68	−0.01	0.02	0.04	0.02	0.05	0.04	0.16
Lactate	X-ray	0.41	0.06	0.11	0.01	0.02	0.05	0.03	0.31
dehydrogenase	CD	0.39	0.07	0.12	0.05	0.01	0.03	0.04	0.29
Papain	X-ray	0.28	0.09	0.00	0.05	0.02	0.03	0.04	0.49
	CD	0.26	0.08	0.04	0.05	0.03	0.02	0.04	0.48

[a]Data from J. P. Hennessey and W. C. Johnson (1981).
[b]H, α helix +3/10 helix; A, antiparallel β sheet; P, parallel β sheet; I, II, III, T, various types of β turns [see Hennessey and Johnson (1981)]; O, other structure. The numbers represent the fraction of each structure.

PROBLEMS

*1. Derive Equation (10.3) from the definition of $\theta = \arctan(b/a)$, where b/a is the ratio of the minor to major axes of the ellipse of amplitudes. You may assume that b/a and $A_L - A_R$ are small compared to unity, where $A_L - A_R$ is the difference in the absorbance for left and right circularly polarized light.

2. The following data list the CD maxima and minima for β-lactoglobulin A under two different conditions.

$\lambda_{max,min}$	$[\theta]$ *(deg-cm²/decimole)*
In aqueous buffer, pH 5	
215	−5,500
196	+8,400
In 99% ethanol, 0.01 *M* HCl	
220	−23,000
208	−26,000
192	+51,000

Describe, qualitatively, what happens to β-lactoglobulin when it is transferred into acidic ethanol solution.

3. Nucleosomes, which contain a length of about 150 base pairs of DNA wrapped about a protein (histone) core, have been studied by thermal denaturation, using both

absorbance at 260 nm and CD. The following data were recorded as a function of temperature:

$T (°C)$	Percent Increase in Absorbance at 260 nm	$[\theta]$ (deg-cm/decimole) $\times 10^{-3}$	
		At 273 nm	At 223 nm
20	0	+0.4	
30	0	+0.4	−39
35	0	+0.5	−39
40	0	+0.7	−39
45	0	+0.9	−39
50	0	+1.5	−39
55	2	+1.8	−39
60	6	+2.2	−39
65	8	+2.7	−37
70	12	+3.5	−26
75	26	+4.8	−18
80	33	+4.5	−15
85	37	+4.5	−14

It has been shown that only the DNA absorbs above 250 nm, and that the protein dominates the CD at 223 nm. Describe, in as much detail as possible, what happens to a nucleosome when it is heated. Under the same solvent conditions, free DNA melts at 42°C. Free DNA has a molar residue ellipticity at 273 nm of about 8×10^3 at room temperature. This *decreases* when DNA melts.

4. *Assume* that the Greenfield–Fasman data for synthetic polypeptides shown in Figure 10.16 describe accurately the curves that would describe the contributions of α helix, β sheet, and random coil in a protein. Pick three wavelengths that would discriminate most sensitively between these forms. Show how you might analyze the data for an unknown protein in terms of CD values at these three wavelengths.

REFERENCES

Charney, E.: *The Molecular Basis of Optical Activity*, John Wiley & Sons, Inc., New York, 1979. A current and comprehensive discussion of both theory and application of optical rotation and circular dichroism.

Hennessey, J. P., Jr. and W. C. Johnson: *Biochemistry*, **20**, 1085–1094 (1981). On the analysis of protein secondary structure by circular dichroism.

Kauzmann, W.: *Quantum Chemistry*, Academic Press, Inc., New York, 1957, Chap. 15. A very readable description of scattering and optical rotatory power from a classical electromagnetic point of view. Chapter 16 treats the same material in terms of quantum theory.

Chapter Eleven

X-RAY DIFFRACTION

One of the most common, if unnoticed, examples of diffraction is the pattern of a cross formed by a distant street light seen through a window screen or through a stretched pocket handkerchief. If the screen is sufficiently fine and if the pattern is examined closely, it will be seen to be made up of a rectangular array of spots of light of varying intensities. This is simply the two-dimensional diffraction pattern of the light by the two-dimensional grid of the screen. Each spot can be identified by counting to the right or left and up or down from the center of the pattern and by assigning these two integers as the "indices" (h, k) of the spot.

With a finer mesh screen the pattern will be seen to be expanded, and with a coarser mesh the spots will be closer together. This reciprocal relationship between diffracting object and diffraction pattern is universally valid, and the coordinate system in which the pattern is measured is referred to as "reciprocal space." If, instead, one now chooses a screen of the same mesh but with a distinctive pattern such as would be produced by adding a finer wire alongside each coarse one to give a Scotch plaid effect, one finds that the spots in the diffraction pattern are in the same places but that their relative brightness has been changed. These experiments demonstrate two fundamental facts about diffraction: The size of the mesh determines the positions of the spots but the nature of the weave determines their intensities. . . . [R. E. Dickerson (1964).]

In this chapter, we shall be concerned with the kind of information that can be obtained from examination of the "mesh and weave" of crystals of macro-molecules.

Probably no other physical technique has had more impact, actual or potential, on biochemistry than X-ray diffraction. The results of these applications in one area alone, the study of globular proteins, provides a startling example. In a few decades, we have progressed from a situation in which these molecules were only vaguely visualized, and experimental studies were literally probing in the dark, to a position in which we *know*, for hundreds of proteins, the precise spatial location of all of the residues in the chain. Now, and only now, studies of the active sites of enzymes become meaningful, the significance of evolutionary changes can be discussed intelligently, and the interactions of protein molecules with other macromolecular structures can be visualized.

In discussing X-ray diffraction, we shall adopt an approach a little different from that which we have used in earlier chapters. X-ray diffraction is a complex and difficult technique, and it is unlikely that the average reader will employ it, as he or she might use sedimentation or light scattering. Even to provide a comprehensive picture of the theory would require more space than we can allow. We shall be content to define the general problems, indicate the limitations and difficulties in the experimental technique, and illustrate the results by a few examples. A fuller description of the theory and methods may be found in some of the references given at the end of this chapter.

11.1 Fundamentals of X-Ray Diffraction

Most elementary discussions of X-ray diffraction begin with a "derivation" of the Bragg law. A crystal is pictured as a set of reflecting planes (Figure 11.1) with uniform spacing from which X rays, incident at an angle θ, are reflected at an angle θ. It is then argued that for reinforcement of the reflected waves to

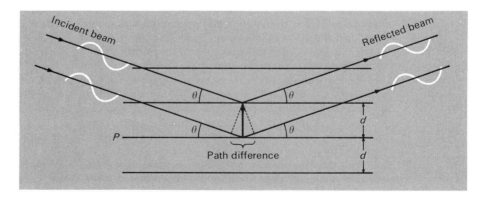

Figure 11.1 A conventional derivation of the Bragg equation. The crystal planes *P* are considered to be reflectors. Reinforcement in the reflected beam will occur only when the path difference is some multiple of the wavelength, λ, so θ must satisfy the equation $n\lambda = 2d \sin \theta$.

occur, the angle θ must satisfy the condition

$$2d \sin \theta = n\lambda \tag{11.1}$$

where n is an integer. If the crystal is large and perfect, containing many parallel planes, destructive interference will result at all angles not stringently satisfying this condition.

Even a glance at such a derivation should provoke skepticism in an alert student. What in the world is doing the reflecting? Are not the "reflecting planes" really made up of layers of atoms or molecules, with more empty space than matter? These objections are quite valid, and yet the Bragg equation is correct. Let us see how a rational derivation can be arrived at.

In the first place, X-ray diffraction is a scattering phenomenon. Its unique characteristics arise because (1) the scattering is from a periodic structure, and (2) the wavelength (λ) of the X rays is comparable to the periodic spacing of the structure. In Chapter 9 we saw that the scattering from a periodic structure with spacings much smaller than the wavelength was largely eliminated by internal interference. If the wavelength and spacing are comparable, interference and reinforcement are both possible. Let us consider X-ray scattering from the simplest periodic structure, a one-dimensional row of scattering centers (atoms, for example) with uniform spacing (c) (Figure 11.2). The direction of incident

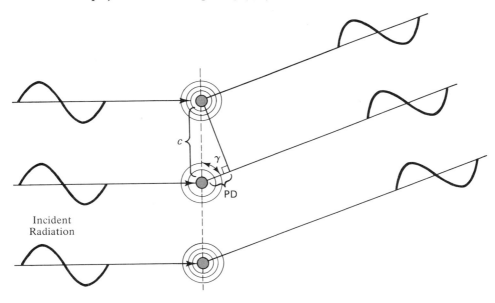

Figure 11.2 Demonstration of the von Laue condition for radiation incident perpendicular to a row of scatterers with regular spacing, c. In scattering directions where the path difference equals a multiple of λ, all scatterers will reinforce. At any other angles, there will be interference, which will become more complete the longer the row of scatterers.

radiation is taken to be perpendicular to the row of atoms. If one now imagines scattering from the centers, this scattering will be in all directions, as is light scattering. However, only in certain directions will reinforcement of scattered rays occur; those will be such that the path difference (PD) between rays scattered by adjacent atoms corresponds to an integral number of wavelengths. If γ is the angle between the direction of the scattered radiation and the row of scatterers, this condition is given by

$$l\lambda = c \cos \gamma \tag{11 2}$$

where l is an integer. If the row of atoms is very long, complete annihilation will occur at all other angles. Then, from such a row, radiation will be scattered only along conical surfaces, as shown in Figure 11.3. A screen or photographic plate

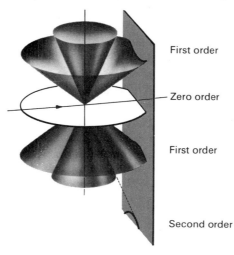

First order

Zero order

First order

Second order

Figure 11.3 The cones along which constructive interference can occur for a row of scatterers. These are shown intersecting a photographic plate. On the plate, layer lines corresponding to $l = 0, 1, 2$, and so forth will appear.

placed perpendicular to the incident radiation will intersect these cones on hyperbolic lines. The spacing of these *layer lines* on the plate will be nearly proportional to $\cos \gamma$, and hence to $1/c$. This is our first encounter with the reciprocal space concept; closer spacings in a crystal correspond to large spacings on the diffraction pattern, and vice versa.

If the incident radiation makes an angle γ_0 other than 90° with the row of scatterers, Equation (11.2) must be modified[†] to

$$l\lambda = c (\cos \gamma - \cos \gamma_0) \tag{11.3}$$

If a two-dimensional array is considered, with spacing a and c in the x and z directions, respectively, two equations must be satisfied simultaneously:

$$l\lambda = c (\cos \gamma - \cos \gamma_0) \tag{11.4}$$

$$h\lambda = a (\cos \alpha - \cos \alpha_0) \tag{11.5}$$

[†]Proof of this is not difficult. It is left to the reader, to promote a better understanding of the equations.

Physically, this means that reinforcement will now occur only along the lines where the cones of diffraction from the x rows intersect those from the z rows (see Figure 11.4). Spots will appear on a photographic plate placed parallel to the two-dimensional array. This pattern is the kind referred to in the quotation from Dickerson.

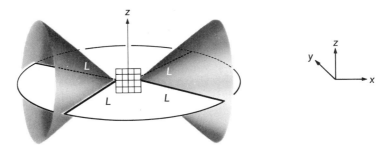

Figure 11.4 Diffraction by a rectangular lattice in the xz plane. For clarity, only one set of cones of diffraction produced by the x spacings is shown, and only the circle (zero-order reflection) from the z axis is shown. Diffraction will be possible along the lines (L) where these surfaces intersect. These could be called the 1, 0 reflections.

If a three-dimensional array, such as an orthorhombic crystal, with spacings a, b, and c in the x, y, and z directions is employed, a third equation must also be satisfied:

$$k\lambda = b \,(\cos \beta - \cos \beta_0) \tag{11.6}$$

These three equations [(11.4) through (11.6)], are called the von Laue equations.

For arbitrary orientations of the crystal, it is not possible to satisfy the von Laue equations simultaneously.† Only when the crystal takes certain orientations with respect to the incident beam can reinforcement occur. Suppose that the crystal is arbitrarily oriented with respect to the beam, and the following question is asked: If we consider an angle 2θ with respect to the incident beam, is there reinforcement in this direction? If so, the von Laue equations must be satisfied. To simplify the nomenclature, we write α for $\cos \alpha$, β_0 for $\cos \beta_0$, and so forth. Note that these are also the *direction cosines* for the diffracted and incident beam with respect to the crystal axes. For direction cosines of a line the following relations apply:

$$\alpha^2 + \beta^2 + \gamma^2 = 1 \tag{11.7}$$

$$\alpha_0^2 + \beta_0^2 + \gamma_0^2 = 1 \tag{11.8}$$

and if the one line is making an angle 2θ with respect to the projection of the other, we have

$$\cos 2\theta = \alpha\alpha_0 + \beta\beta_0 + \gamma\gamma_0 \tag{11.9}$$

---

†This is not difficult to see geometrically. While two nonparallel surfaces will always intersect in a line or lines, three surfaces will mutually intersect along a single line only in special circumstances.

These are simply standard geometric relationships. If we square each of the von Laue equations, we find

$$\frac{h^2\lambda^2}{a^2} = \underline{\alpha}^2 - 2\underline{\alpha}\underline{\alpha}_0 + \underline{\alpha}_0^2 \tag{11.10}$$

$$\frac{k^2\lambda^2}{b^2} = \underline{\beta}^2 - 2\underline{\beta}\underline{\beta}_0 + \underline{\beta}_0^2 \tag{11.11}$$

$$\frac{l^2\lambda^2}{c^2} = \underline{\gamma}^2 - 2\underline{\gamma}\underline{\gamma}_0 + \underline{\gamma}_0^2 \tag{11.12}$$

or, summing,

$$\left(\frac{h^2}{a^2} + \frac{k^2}{b^2} + \frac{l^2}{c^2}\right)\lambda^2 = 1 - 2(\underline{\alpha}\underline{\alpha}_0 + \underline{\beta}\underline{\beta}_0 + \underline{\gamma}\underline{\gamma}_0) + 1$$

$$= 2(1 - \cos 2\theta) \tag{11.13}$$

By a standard identity, $1 - \cos 2\theta = 2 \sin^2 \theta$; therefore,

$$\lambda^2 \left(\frac{h^2}{a^2} + \frac{k^2}{b^2} + \frac{l^2}{c^2}\right) = 4 \sin^2 \theta$$

or

$$\lambda \left(\frac{h^2}{a^2} + \frac{k^2}{b^2} + \frac{l^2}{c^2}\right)^{1/2} = 2 \sin \theta \tag{11.14}$$

This is simply a generalization of the Bragg law [Equation (11.1)] to an ortho-rhombic crystal with spacings a, b, and c.†

Thus the Bragg law is proved, starting with the concept of diffraction as a scattering phenomena. We have *not* shown, however, that the angle θ is in fact a reflection angle from some set of crystal planes. Such is demonstrated, for example, in Lipson and Cochran (1953).

The fact that arbitrary orientations do not allow satisfaction of the von Laue equations requires that we examine a crystal over a range of orientations to observe its diffraction pattern. One common way of doing this is called the *rotating crystal* method. In Figure 11.5 is shown a set of unit cells with indices for some of the sets of crystal planes shown. These are the Miller indices: The 010 planes are perpendicular to the y axis. If the crystal is oriented as shown in Figure 11.6 and then rotated about the z axis, diffraction from the 010 planes will be possible only when these planes make an angle θ with the direction of the beam; this angle must be one that satisfies the Bragg equation for this set of planes.‡ A corresponding spot will appear on the photographic plate. Other spots will appear on this layer line for higher orders of diffraction ($h = 2, 3$,

---

†For example, for 100 planes, we obtain $h\lambda/a = 2 \sin \theta$. The expression in brackets in Equation (11.14) will, in general, represent the reciprocal of the distance between planes with Miller indices h, k, l. If the crystal is not orthorhombic, the expression corresponding to Equation (11.14) is more complicated.

‡At this point it would be very useful for the reader to show that the individual von Laue equations are all satisfied in the orientation shown in Figure 11.6.

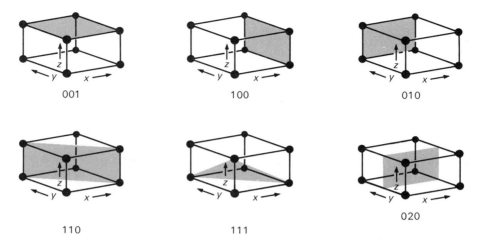

001   100   010

110   111

020

Figure 11.5 Miller indices of some of the possible planes in an orthorhombic unit cell (a rectangular parallelepiped) with sides a, b, c. The numbering rule is as follows: If the plane cuts an axis at a/h, b/k, c/l, the indices are h, k, l.

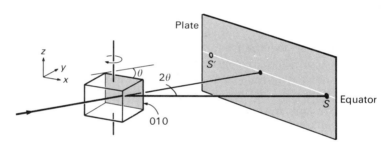

Figure 11.6 The rotating crystal method. The crystal has been oriented with the z axis perpendicular to the beam and is rotated about this axis. When the 010 planes make the Bragg angle θ with respect to the beam, a diffraction spot (S) is produced on the equator of the image at an angle 2θ from the incident beam. Another spot will be observed at S', when the angle is $-\theta$. Higher-order reflections will also be seen. Higher-order spots will be farther out on the pattern.

and so forth) and for higher-order spacings (020 planes and the like). Note that in keeping with the reciprocal space concept, those spots corresponding to larger spacings will appear closer to the center of the pattern. The generation of spots on other layer lines is depicted by Figure 11.7. Here, the 101 planes are so oriented as to satisfy the Bragg condition. In this way, rotation of the crystal in the beam will serve to produce an image that contains many of the diffraction maxima possible for the structure.

It must be reemphasized at this point that we are only touching the fringes of a vast subject matter. For example, the rotating crystal method is by no means the only, or even the most common method for data collection. Most data are

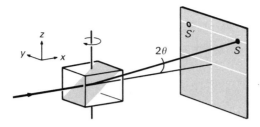

Figure 11.7 Like Figure 11.6 but showing how off-equator spots are given by planes not parallel to the *z* axis.

presently obtained using either a *precession camera*, in which both crystal and film are moved in a synchronized way so as to optimize collection, or by automated diffractometers, in which an X-ray detector and the sample are moved so as to measure the intensity of each spot. For the sake of simplicity, we have discussed only orthorhombic crystal lattices, in which all three crystal axes are mutually perpendicular. Real crystals exhibit a much wider range of lattice types, and one of the first tasks of the crystallographer is to determine into which of these types the particular lattice falls. But to explain such aspects of the problem is beyond the aims of this section, which are to give the reader a general idea of the way in which macromolecular structure can be deduced from crystallographic studies.

Returning to our simple examples, it is clear that the dimensions *a*, *b*, and *c* may be deduced from the spacing of the layer lines and spots on the diffraction pattern. These dimensions define the *unit* cell, a concept central to crystallographic analysis. The unit cell is the minimal entity, which by translation alone, can produce the entire crystal. The unit cell may contain one or more *asymmetric units*, which in turn may each correspond to one or more molecules. The simplest example to think of is that in which a single molecule makes up the asymmetric unit, and there is one such unit per unit cell. An orthorhombic crystal can then be generated by successive displacements of this molecule,

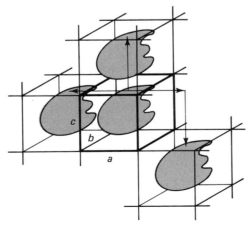

Figure 11.8 Generating an orthorhombic crystal by displacing the unit cell contents by multiples of distances *a*, *b*, and *c* in the *x*, *y*, and *z* directions.

without rotation, by multiples of the distances $a, b,$ and c in the $x, y,$ and z directions. Note that it does not matter where we imagine placing the molecule within the unit cell, the same crystal is generated wherever we place the origin of the cell axes with respect to the molecule (see Figure 11.8).

11.2 Diffraction by Crystals of Macromolecules

The principal aim of crystallography is not the determination of lattice type and unit cell dimensions, although these are necessary preliminaries. It is, rather, deduction of the distribution of matter *within* the unit cell; that is, determination of the three-dimensional structures of the molecules contained therein. To reiterate Dickerson's analogy—we are concerned with the weave of the fabric, rather than its mesh.

For a long time such an objective seemed preposterous for molecules as complex as globular proteins. But the problem has been solved, and the magnitude of the achievement can be appreciated by examining the structure of even a small protein such as cytochrome c (Figure 11.9). The polypeptide chain con-

(a)   (b)

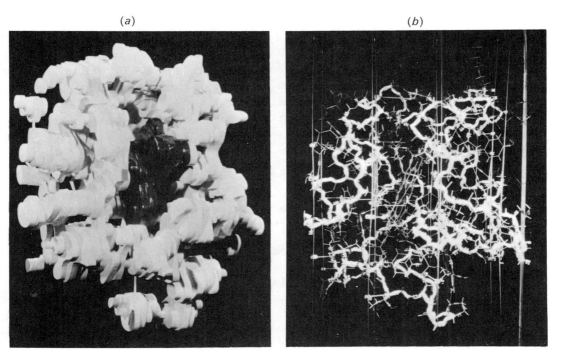

Figure 11.9 Models of cytochrome c at resolution of (*a*) 4 Å and (*b*) 2.8 Å. In the former the chain can barely be traced. In the latter it is possible to see the orientation of individual amino acid residues. Courtesy of R. E. Dickerson.

tains hundreds of atoms, and is folded in a complex manner. The enormous number of reflections observed in the diffraction pattern attests to the regularity of conformation of molecules in the crystal lattice (Figure 11.10).† To deduce

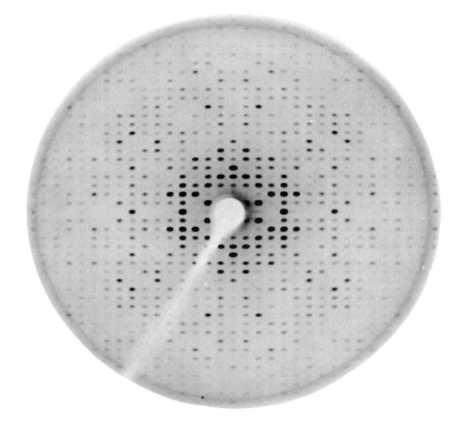

Figure 11.10 X-ray diffraction pattern from a crystal of horse heart oxidized cytochrome *c*. Courtesy of R. E. Dickerson.

the detailed structure of such a molecule, it is clearly necessary to use as much as possible of the information available from the diffraction photograph. This means not only the spacings but also the intensities of the spots. So we must ask, what determines the relative intensities of the reflections? As before, we shall use a simple model to examine the general principles involved.

Consider a unit cell, which we shall assume contains one protein molecule. This is an arrangement of atoms (scattering centers) that are separated by

†The regularity is sometimes astonishing. Good data on myoglobin have been attained to a resolution of 1.4 Å. This means that the great majority of the atoms in a great majority of the molecules in the crystals are held with at least this precision.

distances comparable to the wavelength of the X rays. We can define a set of crystallographic planes by picking the corresponding atom in each molecule and passing the planes through them. Since there is only one molecule per cell, these planes define the unit cell. Any other set of corresponding atoms lies on a similar but displaced set of crystallographic planes. There will be interference between the scattering between these sets of planes. That is, while all sets will give exactly the same set of reflections, their contributions to these reflections will differ in *phase*.

To make this argument even more concrete, consider Figure 11.11, which depicts the simplest possible molecular crystal—one made up of diatomic molecules. Let us assume that all the molecules lie in planes parallel to the *xy*

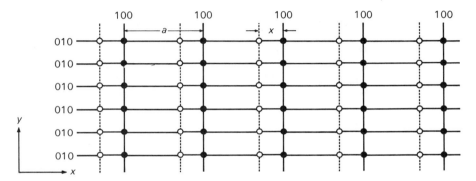

Figure 11.11 The *xy* plane of a hypothetical lattice of diatomic molecules. Note that the same lattice (displaced) may be used either for the white atoms or the black atoms. The molecules are assumed to lie in the plane of the paper.

plane and oriented as shown in the figure. Suppose that the crystal is oriented in the X-ray beam so that the Bragg condition [Equation (11.1)] is satisfied for the 100 planes. We draw a set of such planes (perpendicular to the paper) through the *black* atoms. Note that if the Bragg (and hence von Laue) equations are satisfied, these atoms are scattering in phase in the direction of the reflected beam. This means the phase difference between planes of black atoms is some multiple of 2π rad (360°). Note, also, that the *white* atoms are doing precisely the same thing and are in phase with one another. *But the scattering from the white atoms is out of phase with that from the black.* How much? If the *spacing* in the *x* direction is *a*, and the displacement of the white lattice from the black lattice is *x*, the phase difference must be

$$\Delta\phi = 2\pi\frac{x}{a} \tag{11.15}$$

for the first-order reflections. For higher orders, we have $2\pi hx/a$, where $h = 2, 3, \ldots$. For reflection from any arbitrary set of planes, with indices h, k, l, we have

$$\Delta\phi = 2\pi\left(\frac{hx}{a} + \frac{ky}{b} + \frac{lz}{c}\right) \tag{11.16}$$

where one atom has been chosen to be at the origin of the unit cell, $x = 0$, $y = 0$, $z = 0$. In Chapter 9 we considered the problem of calculating the amplitude and intensity of the radiation scattered from a pair of oscillators. Here we are dealing with a generalization of that problem, the scattering from a group of atoms that make up the contents of the unit cell. While the calculation could be done in terms of sines and cosines, as in Chapter 9, this becomes very unwieldy if many waves of different amplitudes (from centers of differing scattering power) are to be added. The problem becomes much simpler when waves are written in terms of complex numbers. We may represent a periodic displacement of amplitude f_k and phase $\Delta\phi_k$ either by

$$f_k \cos(2\pi\nu t + \Delta\phi_k) \qquad \text{or} \qquad f_k e^{i(2\pi\nu t + \Delta\phi_k)}$$

where $i = \sqrt{-1}$. The representations are physically equivalent, for since $e^{i\theta} = \cos\theta + i\sin\theta$, the real part of the expression to the right (which is the physically meaningful part) is identical to the expression on the left. Thus, if we have a jumble of waves at the point of observation, the total field will be

$$
\begin{aligned}
E &= \sum_k f_k e^{i(2\pi\nu t + \Delta\phi_k)} \\
&= \sum_k f_k e^{i2\pi\nu t} e^{i\Delta\phi_k} = e^{i2\pi\nu t} \sum_k f_k e^{i\Delta\phi_k}
\end{aligned} \tag{11.17}
$$

The amplitude of the sum of a number of waves of different phase and amplitude is seen, from Equation (11.17), to be

$$F = \sum f_k e^{i\Delta\phi_k} \tag{11.18}$$

If the scattering centers are atoms, the f_k's will be proportional to the individual scattering powers of these atoms. If the scattering centers are taken to be small volume elements in the unit cell, the f_k's are proportional to the electron densities in these elements. Returning now to the calculation of the scattering from a lattice of black and white atoms, we may write immediately

$$F_{100} = f_B e^{i\Delta\phi_B} + f_W e^{i\Delta\phi_W} \tag{11.19}$$

where the subscripts B and W refer to black and white atoms, respectively. F_{100} is called the *structure factor* for the 100 reflection. Since we have taken the black lattice as centered, we have

$$
\begin{aligned}
F_{100} &= f_B e^{2\pi i(0/a)} + f_W e^{2\pi i(x/a)} \\
&= f_B + f_W e^{2\pi i x/a}
\end{aligned} \tag{11.20}
$$

Now, the intensity of the reflection will be proportional to the square of the amplitude (or to the product of the amplitude and its complex conjugate,† if

---

†The conjugate of a complex number is the number with i replaced by $-i$. Thus in Equation (11.20) $F_{100}^{*} = f_B + f_W e^{-2\pi i x/a}$.

F is complex):

$$I(hkl) = F(hkl)F^*(hkl) \qquad (11.21)$$

or

$$I(hkl) = F^2(hkl) \qquad \text{if } F \text{ is real}$$

In our example,

$$I(100) = f_B^2 + f_B f_W(e^{2\pi i x/a} + e^{-2\pi i x/a}) + f_W^2 \qquad (11.22)$$

If we use the identity $e^{\pm i\theta} = \cos\theta \pm i\sin\theta$, we can express the middle term as a cosine:

$$I(100) = f_B^2 + 2f_B f_W \cos\frac{2\pi x}{a} + f_W^2 \qquad (11.23)$$

If $x/a = 0$ or 1 (atoms superimposed), we have $I(100) = (f_B + f_W)^2$. Otherwise, the result is smaller; there is destructive interference. In fact, in the special case where $x/a = \frac{1}{2}$, we obtain $\cos 2\pi x/a = -1$, and $I(100) = (f_B - f_W)^2$. If $f_B = f_W$ (the atoms have the same scattering power), this would mean that the intensity of the 100 reflection would be zero. This special case can easily be seen another way; the atoms would simply cancel out one another's scattering in this reflection. The result is thus seen to be a general case of the example in Chapter 9.

This lengthy discussion of a simple problem points out the complexities facing the protein crystallographer. Each reflection can involve thousands of contributions from the atoms in the unit cell, each differing in phase. Equation (11.18) can be generalized to

$$F(hkl) = \iiint_{\text{unit cell}} \rho(x, y, z)e^{2\pi i[(hx/a) + (ky/b) + (lz/c)]} \, dV \qquad (11.24)$$

where $\rho(x, y, z)$ is the electron density in the volume element dV in the unit cell. The situation would be hopeless but for one fact. The crystal is a periodic structure; therefore, $\rho(x, y, z)$ is a periodic function—it must repeat from cell to cell. Equation (11.24) will be recognized by any student acquainted with Fourier series. It is simply the general expression for the coefficients in a Fourier series representation of the periodic function $\rho(x, y, z)$. Thus the electron density may be written as

$$\rho(x, y, z) = \frac{1}{V} \sum_{h=-\infty}^{\infty} \sum_{k=-\infty}^{\infty} \sum_{l=-\infty}^{\infty} F(hkl)e^{-2\pi i[(hx/a) + (ky/b) + (lz/c)]} \qquad (11.25)$$

where V is the volume of the unit cell. Thus, if we knew the $F(hkl)$'s (the *structure factors*), we would be able to calculate $\rho(x, y, z)$ directly and obtain a picture of the unit cell contents (the molecule!). But there is a problem. We can only measure intensities, and these give at best the quantity $F(hkl)F^*(hkl)$. That is, we know the magnitude of a complex quantity but not its phase. The problem can be expressed succinctly by recalling another way of representing a complex number—by a vector on the complex plane (Figure 11.12). We know the length of the vector but not its direction.

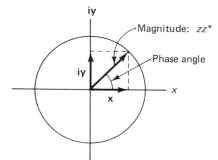

Figure 11.12 A complex number represented as a vector in the complex plane. Note that it is indeed the vector sum of vectors x and *iy* on the axes.

Solutions to this *phase problem* take two general forms. For simple molecules, one can guess at the structure. This gives trial $F(hkl)$'s complete with phase, from which a diffraction pattern can be computed. Comparison of this with the observed pattern may suggest refinements. This procedure is out of the question with protein molecules, but it has been useful both with small biologically important structures such as vitamins and in refining the structures of fibrous biopolymers. A second general approach is known as the *isomorphous replacement* method. This involves adding heavy atoms (strong scatterers) to the protein molecule and investigating their effect on the diffraction. The heavy atoms must not modify either the protein conformation or the crystal form; the replacement must be isomorphous.

Now it is not too difficult to locate one or more heavy atoms in the unit cell (see the Patterson method, below). If the heavy atoms are located, the amplitude and phase of their scattering are known.

The structure factor for the isomorphous derivative containing the heavy atoms $[F_H(hkl)]$ must be the sum of the structure factor for the protein itself $[F_P(hkl)]$ and that for the heavy atoms alone $[F_h(hkl)]$:

$$F_H(hkl) = F_P(hkl) + F_h(hkl) \qquad (11.26)$$

Now of these three complex quantities, we know the magnitudes of the first two and the magnitude *and* phase of the third. If we represent them as vectors, then Equation (11.26) corresponds to a vector sum. If we now draw F_h as a vector, in the imaginary plane (Figure 11.13) F_P and F_H must be vectors which can be drawn from the head and tail of F_h, respectively, and which must meet at a point. Thus, we draw circles, as in Figure 11.13, with radii F_P and F_H. At their intersections, Equation (11.26) will be satisfied, and the phases ae well as the magnitudes will be known. Note, however, that there will be two intersections and two solutions. To choose which is correct, at least one more isomorphous derivative is needed, to provide a third circle, which should pass through one of the two intersections—the correct one. Note also that the F_h should not be of insignificant magnitude as compared to F_P, or the circles will nearly superimpose, and it will be difficult to get an accurate reading of their intersections. This is why

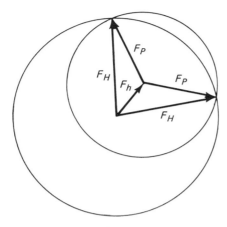

Figure 11.13 Solving for the phase of F_p by isomorphous replacement.

heavy atoms are needed, and is one reason why working with very large protein molecules (or, more precisely, large unit cells) presents difficulty.

A useful technique for the determination of heavy atom positions involves what are known as Patterson maps. The Patterson function is written as

$$P(hkl) = \frac{1}{V} \sum_{h=-\infty}^{\infty} \sum_{k=-\infty}^{\infty} \sum_{l=-\infty}^{\infty} F^2(hkl)e^{-2\pi i(hx/a + ky/b + lz/c)} \qquad (11.27)$$

Note that this is not too difficult to calculate; it contains only the squares of structure factors, which are given by the intensities of the various spots. The Patterson map turns out to be a measure of densities of *distances* between groups. It is most easily visualized by considering a unit cell containing three scatterers [Figure 11.14(*a*)]. If we measure *all* vectors between these centers,

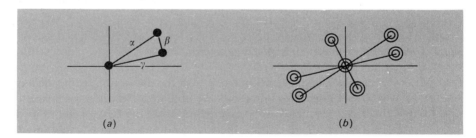

(a)   (b)

Figure 11.14 A planar three-atom "molecule" (*a*) and its corresponding Patterson function (*b*).

move the tails of these vectors to the origin, and graph the positions of their heads, we will have the resulting Patterson map [Figure 11.14(*b*)]. Note that it is more complex than the model itself. For proteins, then, the complete Patterson is almost useless. But a "difference Patterson," obtained by comparison of the diffraction patterns of native and heavy-atom-substituted protein, is much simpler and is of great aid in locating the heavy atoms in the unit cell.

In brief, then, the procedure for deducing the three-dimensional structure of a protein proceeds somewhat as follows:

1. Good crystals of the protein must be prepared. Preliminary studies yield crystal form and unit cell.

2. Several isomorphous replacements must be obtained.

3. Diffraction patterns of a number of these must be obtained and measured.

4. Heavy atom positions must be deduced, and the phases of reflections must be calculated.

5. A trial electron density map is obtained. This may be used to calculate better phases and refine the structure.

At this point, one aspect of the reciprocal space concept must be reiterated. The reflections closest to the X-ray beam correspond to the largest regular spacings; those farther out yield finer details. It can be put another way: Those spots with smallest indices (*hkl*) yield the first coefficients in the Fourier series (11.25). Thus a first approximation to the structure can be obtained by choosing only a relatively small number of reflections (a few hundred on each isomorphous derivative). Further refinement can then be made by taking into account more reflections. At each point, one can speak of the "resolution" of the crystal structure. The models of cytochrome *c* shown in Figure 11.9 illustrate the effect of enhanced resolution. This process cannot be continued indefinitely; for one thing, it gets expensive, the number of reflections needed going up roughly as the inverse square of the resolution. Furthermore, the diffraction patterns do not yield infinite detail. They fade out for small spacings. perhaps because the protein structure itself exhibits variation below certain limits (see Fig. 11.10). However, that this detail may be quite fine is shown by the resolution obtained in Figure 11.9.

11.3 Diffraction by Fibrous Macromolecules

So far, we have been concerned with the diffraction from single crystals, turned so that Bragg reflections could be obtained. But there is obviously another way in which a sample can be assured to satisfy the Bragg conditions. If, instead of a single crystal, we utilize a powder made up of many small, randomly oriented crystals, some will surely be so oriented as to satisfy the Bragg condition for each set of crystal planes. But the crystals will also be randomly oriented with respect to the axis defined by the incident beam. This will smear each spot on the diffraction pattern into a circle about the origin (imagine rotating the crystal in Figure 11.6 about the beam axis). Thus such a *powder pattern*, while giving the lattice spacings, does not allow us to identify them in the way that they can be from a single-crystal photograph (see Figure 11.15). A situation somewhat between that of a single crystal and a powder is found in the X-ray diffraction

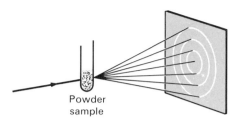

Figure 11.15 The powder method.

of fibers. A fiber may possess some regularity of structure along the fiber axis, with usually less regularity in directions perpendicular to the axis. Since individual crystallites in the fiber will be more or less parallel to the fiber axis, but with various orientations *about* this axis, the same situation is achieved as by rotation of a single crystal. Depending on the extent of orientation in the fiber, the diffraction pattern will exhibit rings, arcs, or spots (see Figure 11.16).

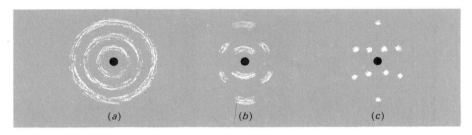

(a)   (b)   (c)

Figure 11.16 Diffraction from a fiber in which molecules are (*a*) unoriented, (*b*) partly oriented, and (*c*) highly oriented.

Most fibrous macromolecules exhibit helical structures, and elongated helices are quite easily oriented by stretching of the fibers. Much information about the structures of polynucleotide, polypeptide, and polysaccharide helices has been deduced from fiber diffraction patterns. A helix is characterized by two basic parameters: the *pitch* or the distance in the direction of the helix axis for one complete turn, and the *radius* of the helix, perpendicular to its axis (see Figure 11.17). A real polymer helix is made up of repeating units (such as the nucleotide pairs in DNA, or the amino acid residues of the α helix) and may have an integral or nonintegral number of residues per turn.

Just as in the row-of-scatterers model depicted in Figure 11.2, a helical molecule will exhibit scattering along a series of layer lines. The spacing between the layer lines will be inversely proportional to the basic repeat of the helix along its axis. If we consider a simple case, in which the helix has an integral number of residues per turn, this basic repeat will be just the pitch of the helix; if one starts at a particular atom or group and proceeds in the helix-axis direction for the distance of the pitch, an identical scattering unit will again be encountered.

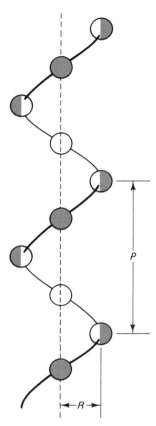

Figure 11.17 A helix with four residues per turn. The pitch is denoted by *P*, the radius by *R*.

For purposes of describing the diffraction pattern from a fiber, imagine the X-ray film to be wrapped cylindrically about the fiber axis (Figure 11.18). The layer lines will then be perpendicular to the fiber axis. The direction on the film parallel to the fiber axis and passing through the X-ray beam is referred to as the *meridian*; the perpendicular to this is the *equator*. The spots of the diffraction pattern will be along layer lines which cut the meridian at spacings proportional to $1/p$. The locations of these spots, and their relative intensities, will depend on the radius of the helix, the number of residues per turn, and details of the helical structure. However, a helical polymer will always yield a characteristic X-like pattern (see Figures 11.18 and 11.19). If there are an integral number, *n*, of residues per turn, the whole pattern will repeat after *n* layer lines. In particular, this means that a reflection will be seen on the meridian at the *n*th layer line. Thus, for B-form DNA, which has 10 base pairs per turn, a strong meridional reflection is seen on the tenth layer line (see Figure 11.19). Spots on the equatorial line of the pattern (the zeroth layer line) are determined by the side-to-side arrangement of the helices in the fiber.

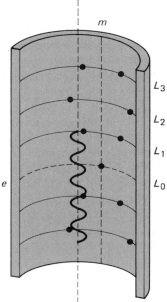

Figure 11.18 Arrangement and nomenclature for fiber diffraction. The meridian and equator are designated by m and e, respectively.

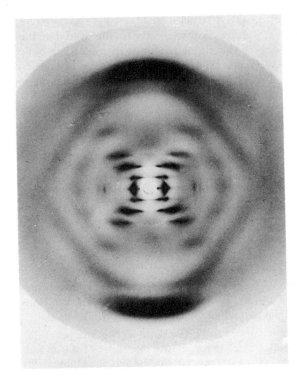

Figure 11.19 X-ray diffraction pattern from wet DNA fiber. Courtesy of R. E. Dickerson.

It is not possible to calculate the electron density distribution directly from such a fiber pattern, as can be done from single-crystal diffraction. Basically, this is because the individual helices in a fiber, even if parallel, will generally be rotated randomly about their axes with respect to one another. Thus the average electron density *about* the fiber axis is "smeared." Analysis of fiber diffraction patterns proceeds in a less direct way. From the pattern itself, the pitch, the number of residues per turn, and the average helix radius can be deduced. Models are then proposed, based on these data, the known bond angles and distances in the polymer, and sterically reasonable guesses as to rotations about individual bonds. From such a model, it is obviously possible to calculate structure factors, and hence the intensities to be expected for the individual spots on the patterns. Comparison with the observed pattern suggests successive refinements in the pattern until a best fit is obtained. The structures of many fibrous polynucleotides, polypeptides, and polysaccharides have been derived in this way.

It must be emphasized that such structures have not been determined with the rigor that accompanies the solution of the structure of a molecular crystal. For this reason, there have been continuing uncertainties concerning the structures of even such well-studied molecules as DNA. Recently, it has been found possible to form molecular crystals from short, identical DNA molecules. The results of single-crystal analysis of such molecules have tended to confirm, in general, the structures envisioned from fiber studies, but indicate subtle local deviations from those idealized DNA structures.

The application of X-ray diffraction methods to the study of biological macromolecules has been one of the major triumphs of molecular biology. As techniques improve, and experience accumulates, we may expect these investigations of biological structure at the molecular level to continue at an ever-quickening pace.

PROBLEMS

1. Given the volume of a unit cell, V, the weight fraction of water, f_w, and the crystal density, ρ, derive an equation to calculate the number (n) of protein molecules of molecular weight M in one unit cell.

2. The milk protein β-lactoglobulin has a molecular weight of 18,363. It crystallizes in an orthorhombic lattice with dimensions 69.3 Å × 70.4 Å × 156.5 Å. The crystals have a density of 1.14 g/cm³, and contain 46 percent water (by weight). How many molecules are there per unit cell? (See Problem 1.)

3. Consider a face-centered-cubic lattice having identical atoms located at the corners and at the centers of all faces of the cubic unit cell. Which reflections will be missing in X-ray diffraction photography?

4. A body-centered-cubic lattice can be visualized as two interpenetrating cubic lattices (as in the NaCl crystal). Calculate the relative intensities of 100 and 110 reflections if the units in one lattice have twice the scattering amplitude as those in the other.

*5. A pair of scattering centers (A and B) are separated by a distance a, and radiation is incident in a direction perpendicular to the line between them. The centers are to be considered of unequal scattering power. What is the intensity of scattering observe at a scattering angle of θ from the incident beam? At what angle θ (other than $\theta = 0$) will the first maximum in scattering occur?

REFERENCES

Blundell, T. L. and L. N. Johnson: *Protein Crystallography*, Academic Press, Inc., New York, 1976. Lots of detailed information about the practice of protein crystallography.

Dickerson, R. E.: "X-ray Analysis and Protein Structure," in *The Proteins*, 2nd ed. (H. Neurath, ed.), Academic Press, Inc., New York, 1964, Vol. II, pp. 603–778. Although somewhat old, this remains a beautifully clear and useful introduction to the topic.

French, A. D. and K. H. Gardner (eds.): *Fiber Diffraction Methods* (ACS Symposium Series No. 141), American Chemical Society, Washington, D.C., 1980. Not a useful text, but this collection of papers gives a good overview of the current status of fiber diffraction studies.

Lipson, H. and W. Cochran: *The Determination of Crystal Structures*, G. Bell & Sons Ltd., London, 1953. A standard reference for X-ray diffraction analysis.

Sherwood, D.: *Crystals, X-rays and Proteins*, John Wiley & Sons, Inc., New York, 1976. Includes a very complete introduction to the mathematical background to X-ray diffraction theory.

Wing, R., H. Drew, T. Takeno, C. Broka, S. Tanaka, K. Itekura, and R. E. Dickerson: *Nature*, **287**, 755–758 (1980). The first single-crystal study of a DNA molecule.

ANSWERS TO

ODD-NUMBERED PROBLEMS

Chapter 1

1. $q = 1.93$ kcal, $w = -1.93$ kcal, $\Delta E = 0$, $\Delta H = 0$.

3. (a) -0.040 cal/mole.

(b) 5.26 cal/deg-mole.

(c) $-0.008°C$.

5. Given 99 bonds, with three conformations each, $\Delta S = 216$ cal/deg-mole. For $\Delta G = 0$, $\Delta H = 70$ kcal, or 0.7 kcal/residue, which is well within the range.

7. $n_i/n_0 = e^{-Ki^2/RT}$, which is a Gaussian distribution. Taking $n_0 = 1$, $n_1 = 0.849$, $n_2 = 0.522$, $n_4 = 0.074$, $n_{10} = 9 \times 10^{-8}$.

Chapter 2

1. $\mu_2 - \mu_2^0 = RT \ln C_2 + 2BRTM_2C_2$. Note that the standard state is a hypothetical state where $C_2 = 1$, but solution behaves ideally.

3. (a) Note that Equation (2.40) may be written

$$\mu_1 - \mu_1^0 = \frac{-RTV_1^0 C_2}{M_2}(1 + BM_2C_2)$$

(b) 0.86 percent.

(c) 14.4 percent.

5. (a) 1.35×10^5 g/mole.

(b) 0.38×10^5 g/mole. Since $(a)/(b) = 3.6$, there are probably four subunits.

7. (a) $B = 2.31 \times 10^{-5}$ liter/mole/g², of which over 99 percent comes from the Donnan term; $1 + BM_2C = 2.52$.

   (b) 1.03×10^{-6} liter-mole/g². Now only about 90 percent is from the Donnan term; $1 + BM_2C_2 = 1.07$.

9. First, recognize that the osmotic pressure will be given by $\pi = RT(c_1 + c_2)$, where c_1 and c_2 are the *molar* concentrations of monomer and dimer. We have an equilibrium constant $K = c_2/c_1^2$, and the relation $c_1 + 2c_2 = c_0$, where c_0 is the total molar concentration of subunits. We want to express π in terms of c_0. You must solve a quadratic equation to get c_1 and c_2, and then use the series expansion

$$(1 + x)^{1/2} \simeq 1 + \frac{x}{2} - \frac{x^2}{8} \cdots$$

This will yield, if concentration is expressed in weight units C_0,

$$\pi = \frac{RTC_0}{M}\left(1 - \frac{KC_0}{8M} + \cdots\right)$$

which proves the statement, since K and M are positive numbers.

Chapter 3

1. (a) -1.7 kcal/mole.

   (b) 17.69.

   (c) To the left.

3. $n = 4.9$ (probably five sites), $K = 9.7 \times 10^2$ liters/mole. Since a Scatchard plot gives a fairly good straight line, sites are probably identical and independent.

5. (a) One can do this by simply writing the MWC equation for $\bar{v}$ for the case $n = 1$. A bit of algebra then yields the result

$$\bar{v} = \frac{1 + Lc}{1 + L}\alpha \bigg/ \left(1 + \frac{1 + Lc}{1 + L}\alpha\right)$$

where c and α are defined as in the text. This is obviously a noncooperative binding curve, with an apparent k given by

$$k = k_R\frac{1 + Lc}{1 + L}$$

Thus the existence of a conformational equilibrium is *not* sufficient to yield cooperative binding; multiple sites are needed.

   (b) $k = k_R\left(\dfrac{1 + Lc}{1 + L}\right)$

7. (a) Since $n = 2$, $c = 0$, we obtain immediately

$$\bar{v} = \frac{2\alpha}{1 + \alpha + L/(1 + \alpha)}$$

   (b) This simply requires writing $\bar{v}/(2 - \bar{v})$, and calculating.

   (c) This is not difficult, but involves a lot of algebra. One must differentiate $\log[\theta/(1 - \theta)]$ with respect to $\log \alpha$ to get the slope. Then differentiate again, and set the second derivative equal to zero. There will be three solutions at $\alpha = 0$, ∞, and a finite value. The latter gives the value of α for the maximum slope. Inserting this into the first derivative, and doing some more algebra, one eventually obtains

$$n_{\max} = 1 + \frac{\sqrt{1+L}-1}{\sqrt{1+L}+1}$$

which for large values of L becomes very close to 2.

9. (a) Since the first two bonds do not count, we have 3^{n-2} conformations, $\Delta S = (n-2)\,\Delta S_{res}$, where $\Delta S_{res} = R\ln 3 = 2.18$ e.u./mole.
    (b) In calculating ΔH, one must note that an α helix of n residues can form only $n-4$ hydrogen bonds. (Draw a helix, with correct bonding pattern, to see this.) So $\Delta H = (n-4)\,\Delta H_{res}$. So $\Delta G = (n-4)\,\Delta H_{res} - T(n-2)\,\Delta S_{res}$. This gives $\Delta H_{res} = 0.7$ kcal/mole, a not unreasonable value.
    (c) Since $T_m = (n-4)\,\Delta H_{res}/(n-2)\,\Delta S_{res}$, an infinite helix will melt at $T_m^\infty = \Delta H_{res}/\Delta S_{res}$. Therefore, $T_m = [1 - 2/(n-2)]T_m^\infty$ for n residues. Even for fairly long helices an appreciable "end effect" is predicted. This is a rather simpleminded model, since it assumes complete cooperativity.

11. For reaction (1): $\Delta G^0 = -9.5$ kcal/mole, $\Delta H^0 \simeq 0$, $\Delta S^0 \simeq 33$ e.u./mol. For reaction (2): $\Delta G^0 = -10.7$, $\Delta H^0 = 28$, $\Delta S^0 = 131$.

Chapter 4

1. The differential equation can be solved via the substitution $F - fv = u$, followed by integration. The result is

$$v = v_{\max}[1 - e^{-(f/m)t}]$$

where $v_{\max} = F/f$. Substituting in the given numbers, with $\eta = 1\times10^{-2}$, yields $t = 1.5\times10^{-10}$ sec.

3. $\ln\dfrac{C_B - C_A}{(C_B - C_A)_0} = -\dfrac{2DSt}{v\,\Delta x}$.

5. The first substitution should lead you to the equation $d^2C/du^2 = -(u/2D)\,dC/du$. Now substitute again, and integrate. One uses the boundary conditions to yield the helpful equation

$$C_0 = \int_{-\infty}^{\infty} \frac{\partial C}{\partial x}\,dx$$

To get Equation (4.18), integrate (4.19) from $-\infty$ to x. Then note that the integral from $-\infty$ to zero is known.

7. (a) $D_{20,w} = 8.6\times10^{-7}$ cm^2/sec.
    (b) $D_{20,w} = 6.7\times10^{-7}$ cm^2/sec.
    (c) $w = 0.093$ cm.
    (d) $\tau_0 = 6.7\times10^{-5}$ sec.

Chapter 5

1. $M = 6.64\times10^4$ g/mole, $f = 5.90\times10^{-8}$ g/sec, $f_0 = 4.52\times10^{-8}$ g/sec, and $f/f_0 = 1.31$. Then, from Figure 4.2, $a/b \simeq 5.8$; alternatively, hydration $= 1.25$ cm^3 of H_2O per cubic centimeter of protein.

3. The average obtained in *general*, is

$$\overline{M_{s,D}} = \frac{\sum M_i^{1-b}C_i}{\sum M_i^{-b}C_i}$$

When $b = 0$, this becomes a weight average; when $b = 1$, it becomes a number average.

5. (a) $M = 2.16 \times 10^5$ g/mole; probably homogeneous.

   (b) Stokes' radius = 54 Å, $f/f_0 = 1.4$. From these results, nucleosomes appear to be roughly globular particles, with a diameter of about 100 Å.

   (c) No. If the DNA were extended, the particle would be about 500 Å long. Calculating the volume of the particle, we find that this would correspond to a prolate ellipsoid with axial ratio $a/b \simeq 17$. This is much too large to be consistent with the data above. The DNA must be coiled or folded.

7. (a) 65 percent.

   (b) 31 percent.

9. The peak separation is 0.70 mm. But σ for both bands is about 0.37 mm. Resolution will be marginal, at best. Lower speed will not help; both separation and σ will increase by the same factor.

11. (a) $s_2/s_1 = 1.50$ [note that the summation here is $(1/R_{12} + 1/R_{21})$]. Therefore, $s_2 = 6.75S$.

    (b) s_4/s_1(linear) $= 2.08$; s_4/s_1(tet.) $= 2.50$.

Chapter 6

1. $U = 1.49 \times 10^{-5}$ cm²/sec-v.

3. (a) $z_N = -0.13$, $z_{sc} = +0.20$, $\Delta z = 0.33$.

   (b) $z_N = -0.53$, $z_{sc} = +0.81$, $\Delta z = 1.33$.

   A difference of $\Delta z = 2$ would be expected. Equation (6.4) is very bad; even Equation (6.5) is not very good.

5. (a) $v = \dfrac{m(1 - \bar{v}\rho)\omega^2 r}{f} - \dfrac{zeE}{f} = s\omega^2 r - UE.$

   (b) 55.2 V/cm.

   (c) No. The boundary will have finite width because of diffusion. Consider what will happen in regions just centrifugal or centripetal to the boundary center.

7. (a) -6.5×10^{-6}.

   (b) -6.5×10^{-2} [although Equation (6.23) may not be accurate for so large a value of $\mu E/kT$].

   (c) $54.7°$.

9. (a) 333 sec⁻¹.

   (b) 387 sec⁻¹.

Chapter 7

1. (a) $[\eta] = 23.5$ ml/g.

   (b) Since $[\eta] = 3.7$ ml/g for the native protein, this represents an increase far too large to be accounted for by hydration. Clearly, the molecule is unfolded, probably as a result of electrostatic repulsion between protonated groups.

3. For intrinsic viscosity, one would observe a 20 percent increase. On the other hand, the sedimentation coefficient will increase by about 52 percent, making this the better method.

5. (a) $\beta = 2.24 \times 10^6$; this is too large for any oblate ellipsoid, hence prolate is better.

(b) 0.54 cm³ of H_2O per cubic centimeter of dry protein. This is a reasonable value.

(c) f/f_0(predicted) $= 1.44$. Therefore, the data are at least approximately self-consistent.

7. About 22 hr.

Chapter 8

1. $A = 0.06, 0.6$; percent $T = 87$ percent, 25 percent.

3. For each wavelength one can write an equation of the form $A_\lambda/l = C_{A\lambda}\epsilon_A + C_{B\lambda}C_B$. These may be solved, by determinants. The λ values must be chosen so that the determinant in the denominator does not vanish.

5. (a) $A_0 = 0.306$, $\tau = \tau_F \bar{A}_{H_2O}/(A_0 - \bar{A}_{H_2O}) = 14$ nsec.

(b) $v_m = 6.0 \times 10^{-20}$ cm³. From the molecular weight and $\bar{v}$, we obtain $v_m = 2.5 \times 10^{-20}$ cm³. Therefore, unless the protein is quite asymmetric, it must be a dimer.

7. The simplest form is

$$\frac{1}{P} - \frac{1}{3} = \left(\frac{1}{P_0} - \frac{1}{3}\right)\left(1 + \frac{\tau_F k}{v_m}\frac{T}{\eta}\right)$$

Chapter 9

1. Defining an element of surface area, $2\pi r^2 \sin\theta \, d\theta$, over which the scattering intensity is constant, we have

$$\int_0^\pi 2\pi r^2 i \sin\theta \, d\theta = I - I_0$$

Substitution of i from Equation (9.33) and integration gives $I - I_0 = (16\pi/3)R_\theta I_0$. In most cases, $I - I_0 \ll I_0$, so we can use the approximation

$$-\tau = \ln\frac{I}{I_0} = \ln\left(1 - \frac{I - I_0}{I_0}\right) \simeq -\frac{I - I_0}{I_0}$$

so $\tau = (16\pi/3)R_\theta$.

3. Calculating the weight-average molecular weight of the mixture, and taking

$$\text{percent error} = \frac{(M_w - M_p) \times 100}{M_p}$$

where $M_p =$ the protein molecular weight, we obtain about 50 percent error.

5. $i(\text{rod})/i(\text{circle}) = 0.95$. The experiment is possible but no longer the practical way to do it. It is much easier to detect breaking by gel electrophoresis (see Chapter 6).

7. $R_G = 78$ Å, diameter $= 201$ Å.

Chapter 10

1. It is best to start by defining the ellipticity as $\theta = \arctan[(E_R - E_L)/(E_R + E_L)]$. Use the fact that $\arctan x \simeq x$, for small x. Also note that $A_L = \log(I_0/I_L)$, and that the intensities are proportional to E^2. The approximation $\ln(1 + x) \simeq x$ is also needed. You will obtain $\theta = 33(A_L - A_R)$. By dividing by lc, and noting the differences in units commonly used for $[\theta]$ and ϵ, one obtains Equation (10.3).

3. The absorbance change is bimodal, indicating two steps in DNA melting. The protein CD (223 nm) changes only at the second stage. The DNA CD (273 nm) starts to increase before the first melting stage, and increases sharply and then decreases slightly at the highest T. The simplest interpretation is that a part of the DNA adopts a more native conformation, and then melts, in the first step. The remainder of the DNA becomes more native-like as the protein denatures; it is then finally able to melt.

Chapter 11

1. $n = \dfrac{\mathfrak{N} \rho V (1 - f_w)}{M}$.

3. First, you should note that the atoms, as described, are each shared by two or more unit cells. To generate the whole lattice, one need place whole atoms only at the points $(0, 0, 0)$, $(0, \frac{1}{2}, \frac{1}{2})$, $(\frac{1}{2}, 0, \frac{1}{2})$, $(\frac{1}{2}, \frac{1}{2}, 0)$. There are actually only four atoms/unit cell. It is easy to show, then, that the structure factor F_{hkl} can be written

$$F_{hkl} = f[1 + e^{i\pi(k+l)} + e^{i\pi(h+l)} + e^{i\pi(h+k)}]$$

Recalling that $e^{i\pi} = -1$, we find four possibilities:
  1. All h, k, l, odd: $F_{hkl} = 4f$.
  2. All h, k, l, even: $F_{hkl} = 4f$.
  3. Two are odd, one even: $F_{hkl} = 0$.
  4. Two even, one odd: $F_{hkl} = 0$.

5. $I(\theta) = I_A(\theta) + I_B(\theta) + 2\sqrt{I_A(\theta)I_B(\theta)} \cos\left(\dfrac{2\pi a}{\lambda} \sin \theta\right)$. The first maximum (other than $\theta = 0$) will be at $\sin \theta = \lambda/a$; $\theta = \arcsin(\lambda/a)$.

INDEX